A Brief Journey in Discrete Mathematics

Randolph Nelson

A Brief Journey
in Discrete Mathematics

 Springer

Randolph Nelson
(Home address)
Beverly, MA, USA

ISBN 978-3-030-37863-9   ISBN 978-3-030-37861-5   (eBook)
https://doi.org/10.1007/978-3-030-37861-5

Mathematics Subject Classification: 97N70

This Springer imprint is published by the registered company Springer Nature Switzerland
AG.
The registered company address is: Gewerbestrasse 11, 6330 Cham, Switzerland

Beauty is the first test:
There is no permanent place in the world for
ugly mathematics.

G. H. Hardy (1877–1947)

Dedicated to my wife - Cynthia Nelson

Acknowledgments

I would like to thank several of my colleagues at work who shared their enthusiasm and interest for the book over the many years it took to write. Sean McNeill suffered through many monologues concerning mathematical relationships that I found particularly beautiful and I remember Rod Dodson having to undergo similar orations. Eric Norman was always receptive to the ideas in the book and even had his daughter read some chapters. Linda Wu, a fellow mathematician, shared her exuberant interest in the material and continually encouraged me towards publication.

I have sat across a restaurant table literally hundreds of times with my business partner, Ira Leventhal. Frequently, during our conversations, the topic would veer away from business towards mathematics. Invariably, this led me to profit from Ira's uncanny intuition regarding numerical relationships.

I would be amiss without acknowledging two professors who have had a major influence on my research career. Rod Oldehoeft's warm and enthusiastic reception of a new student who walked into his office one day started my career. Leonard Kleinrock, my doctorate thesis advisor, opened my eyes to the joys of mathematical modeling with his unique and charismatic view of the subject which launched me on my own research path. I am forever grateful to these professors for their life changing guidance.

I would also like to thank the editor at Springer, Elizabeth Loew, who offered her encouragement for the book and guided its reviews. The last anonymous reviewer, who I wish to particularly thank, made several cogent criticisms that improved the book.

This book is dedicated to my wife, Cynthia, without whom, in so many ways, it would never have seen the light of day. Appreciation also goes to my children, Austin, Caresse, and Cristina, who sometimes lost their father even while he was physically present.

Contents

Chapter 1
Introduction

> *Suit the action to the word, the word to the action,*
> *with this special observance, that you o'erstep not the*
> *modesty of nature:*
> *for any thing so o'erdone is from the purpose of*
> *playing,*
> *whose end, both at the first and now, was and is, to*
> *hold as*
> *'twere the mirror up to nature:*
> *to show virtue her feature, scorn her own image,*
> *and the very age and body of the time his form and*
> *pressure.*
>
> William Shakespeare (1564–1616)
> Hamlet Act 3, scene 2

Hamlet's advice to actors, quoted above, speaks to the essence of mathematics: the action and word are neither more nor less than what is required, the virtue of pure thought and its ageless body transcends time and, finally, that mathematics is the mirror that reflects the face of nature.

In this brief book, the mirror of mathematics is held up to reflect the light of discrete mathematics. This area, with results that can be typically expressed in terms of integer ratios, forms a core discipline within the expansive field of mathematics. Topics addressed include combinatorics, properties of symmetric functions, the Golden ratio as it leads to k-bonacci numbers, non-intuitive and surprising results found in a simple coin tossing game, analysis of sums of integer powers and triangular numbers, the playful, trick question aspects of modular systems, exploration of basic properties of prime numbers including a proof that a prime always exists between a number and its double, and a derivation of bewildering results that arise from approximating irrational numbers as continued fraction expansions. The Appendix

© Springer Nature Switzerland AG 2020
R. Nelson, *A Brief Journey in Discrete Mathematics*,
https://doi.org/10.1007/978-3-030-37861-5_1

contains the basic tools of mathematics that are used in the text along with a numerous list of identities that are derived in the body of the book.

It might seem surprising that so many areas of analysis fall under the auspices of discrete math but, in fact, many aspects of discrete math are not covered such as set theory, logic and Boolean algebra, algorithms, and the theory of computation, graph theory, and matrix theory. The references found at the back of the book are a good starting point for investigations in these areas.

The approach in the book is concise and proceeds directly from first principles. On one occasion a result that lies outside of the book is lifted from probability theory to expedite a derivation. Otherwise, the book is self-contained and does not rely on results not derived within its pages. A reader should be well versed with symbol manipulation and also have some fluency with algebra.

Each chapter follows a sequence of straightforward questions to a logical end that lies at the doorstep of a field of study. In the writer's opinion, these results are beautiful and often surprisingly non-intuitive. This frames the wonder of mathematics and highlights the complex world that lies behind a series of simple, mathematical, deductions.

Chapter 2
Let Me Count the Ways

> *How do I love thee? Let me count the ways.*
> *I love thee to the depth and breadth and height*
> *My soul can reach, when feeling out of sight*
> *For the ends of being and ideal grace.*
>
> Elizabeth Barrett Browning (1806–1861)

Elizabeth Browning probably didn't realize that she was really talking about mathematics when she penned her 43rd sonnet, *How Do I Love Thee?* This chapter provides a more comprehensive answer to this question than Browning was able to present in the remaining stanzas where she enumerates the ways she loves the veiled object of her sonnet. With the power of mathematics, equations are derived that provide a thorough enumeration, leaving no stone untouched. This is done through the simple expedient of selecting a set of items from a set. It is surprising, as when one falls in love, how fast innocent simplicity explodes into a tangled web of complexity. Perhaps this is what makes love stories, and mathematics, so enduringly interesting.

Assume there are n distinguishable items in a set from which k items are selected.[1] The object of this chapter is to count the number of possible ways to make such a selection. There are four different counting paradigms that depend upon the order items are selected and whether selected items are removed or returned to the set. *Selection with replacement* occurs when selected items are returned, otherwise the selection is termed *selection without replacement*. A *permutation* occurs when the order of selected items is maintained, otherwise the selection corresponds to a *combination*.

---

[1] For example, different colored balls are *distinguishable* whereas electrons which have no discernible differences are *indistinguishable*.

© Springer Nature Switzerland AG 2020
R. Nelson, *A Brief Journey in Discrete Mathematics*,
https://doi.org/10.1007/978-3-030-37861-5_2

To illustrate the four different selection paradigms, consider the case where 2 items are selected from the set $\{1, 2, 3\}$. A selection can be represented by an integer so that 31 corresponds to first selecting item 3 followed by item 1. The first line of the table below shows that there are 6 possible permutations when items are not returned to the set (possibilities represented by $\{12, 21, 13, 31, 23, 32\}$). When order is not maintained (so that the selection 12 is counted as being the same as the selection 21) then the number of possibilities reduces to 3 (possibilities represented by $\{12, 13, 23\}$). When items are returned after selection, the number of possibilities in each of the above cases increases by 3 corresponding to the addition of possibilities given by $\{11, 22, 33\}$.

	#	Permutation	#	Combination
Without Replacement	6	$\{12, 21, 13, 31, 23, 32\}$	3	$\{12, 13, 23\}$
With Replacement	9	$\{12, 21, 13, 31, 23, 32, 11, 22, 33\}$	6	$\{12, 13, 23, 11, 22, 33\}$

The rest of this chapter derives equations for the number of possible selections for each of the four counting paradigms, establishes relationships between them, and derives identities that arise from the resultant equations.[2] Throughout this chapter, let the number of different items in a set be denoted by n from which k items are selected.

2.1 Permutations: With and Without Replacement

We first consider permutations when items are not returned to the set, a quantity that is denoted by $p^r_{k,n}$.[3] This value satisfies the recurrence

$$(2.1) \qquad\qquad p^r_{k,n} = n p^r_{k-1,n}, \quad n \geq 1, \ k \geq 1$$

Initial values of the recurrence are $p^r_{0,n} = 1$ and $p^r_{k,n} = 0$ for $k < 0$ or $n < 0$. To explain (2.1) note that the first of the k selections can be done in n different ways, leaving $k - 1$ items left to be selected. Since

---

[2]See the Appendix for a review of using recurrence to solve problems.
[3]The "r" superscript means *with replacement* rather than being a numeric index value.

the selected item is returned to the step, this leaves $p^r_{k-1,n}$ remaining possibilities. Recurrence (2.1) can be solved to yield

(2.2) $$p^r_{k,n} = n^k$$

Values of $p^r_{k,n}$ for small parameter values are given in the following table:

	Values of $p^r_{k,n} = n^k$							
$k \backslash n$	2	3	4	5	6	7	8	9
1	2	3	4	5	6	7	8	9
2		9	16	25	36	49	64	81
3			64	125	216	343	512	729
4				625	1,296	2,401	4,096	6,561
5					7,776	16,807	32,768	59,049
6						117,649	262,144	531,441
7							2,097,152	4,782,969
8								43,046,721

The number of different possibilities for selecting permutations without replacement, denoted by $p_{k,n}$, satisfies the recursion

(2.3) $$p_{k,n} = np_{k-1,n-1}, \quad n \geq 1, \ k \geq 1$$

with initial values of $p_{0,n} = 1$ and $p_{k,n} = 0$ for $k < 0$ of $n < 0$.[4] To explain this, note again that the first selection can be done in n different ways. Since the selected item now is removed from the set, this leaves the remaining $k-1$ items to be selected from a set of $n-1$ items, thus the quantity $p_{k-1,n-1}$. This recurrence can be solved to yield

(2.4) $$p_{k,n} = n^{\underline{k}}$$

where the *lower factorial* is defined by

(2.5) $$n^{\underline{k}} = n(n-1)\cdots(n-k+1)$$

---

[4]The restrictions to have non-negative arguments for $p^r_{k,n}$ and $p_{k,n}$ can be relaxed but will not be considered in this book.

As an aside, note that two algebraic identities follow directly definition (2.5):

$$(2.6) \qquad n^{\underline{k}} = n^{\underline{\ell}} \, (n - \ell)^{\underline{k-\ell}}, \quad \ell \le k$$

and

$$(2.7) \qquad n \, n^{\underline{k}} = n^{\underline{k+1}} + k n^{\underline{k}}$$

Values of $p_{k,n}$ for small parameter values are given in the following table (notice the size differences between this and the table above):

	\multicolumn{8}{c	}{Values of $p_{k,n} = n^{\underline{k}}$}
$k\backslash n$	2	3	4	5	6	7	8	9
1	2	3	4	5	6	7	8	9
2		6	12	20	30	42	56	72
3			24	60	120	210	336	504
4				120	360	840	1,680	3,024
5					720	2,520	6,720	15,120
6						5,040	20,160	60,480
7							40,320	181,440
8								362,880

So far the analysis is straightforward; however, a counterintuitive result follows from these the two defining equations (2.3) and (2.4). The *birthday problem* is typically stated by calculating the probability that, from a set of k people, at least two people have the same birthday. To calculate this, assume that birthdays occur uniformly throughout the year.[5] If all birthdays are unique, then a selection from 365 days of k items is a permutation without replacement. Setting $n = 365$ in equation (2.4) shows that there are $365^{\underline{k}}$ such permutations. The number of total possible selections from 365 days that allows duplicate days is equivalent to selecting k days with replacement which is given

---

[5] A smoothed plot of the frequency of birthdays ascends from a low around January until it reaches a peak in September. There is thus a greater chance that two or more people have a common birthday than what is calculated from equation (2.8).

by equation (2.2), 365^k. Thus, the fraction of random k selections where there are no duplicate birthdays equals[6]

(2.8)
$$\frac{p_{k,n}}{p_{k,n}^r} = \prod_{j=1}^{k-1}\left(1 - \frac{j}{n}\right)$$

To determine the probability that there are multiple birthdays, subtract (2.8) from 1. The numerical results are surprising as shown in the following table:

k	10	15	20	25	30	35	40	45	50	55
$1 - p_{k,n}/p_{k,n}^r$	.117	.253	.411	.569	.706	.814	.891	.941	.97	.986

The break-even point occurs when $k = 23$, showing that there is a 50.72% probability that two people have the same birthday! The table shows how fast this percentage increases with k so that the 99% level is breached at $k = 57$. A fun way to see mathematics in action is to grab the microphone at a typical wedding and ask people having a birthday today to raise their hands. It will almost never fail that at least a couple of hands shoot up!

2.1.1 Dearrangements

To explain another problem that can be solved only using equations (2.3) and (2.4), consider a list of the integers $(1, 2, \ldots, n)$ that are permuted to a new ordering $(q_1, q_2, \ldots, q_n)$ so that no integer is in its original position, $q_i \neq i$, $i = 1, \ldots, n$. Such a rearrangement is termed a *dearrangement*. Denote the number of such possibilities, by g_n which has boundary values: $g_0 = 1$ and $g_1 = 0$. Consider integer j and assume that after permutation it is in position k, and hence $g_k = j$. There are $n-1$ possibilities to select k if $k \neq j$. The number of remaining dearrangements that are possible depend on where integer k is permuted. If k exchanges position with j, so that $g_j = k$, then this leaves $n - 1$ remaining items in the dearrangement, a value given by

---

[6]The value of $p_{k,n}$ and $p_{k,n}^r$ soon swamps a computer's floating point range for large argument values. This is the reason why the value is computed as the product of simple ratios.

g_{n-1}. If this is not the case and integer k is found in position $\ell \neq j$ so that $g_\ell = k$, then two slots of the permutation are taken leaving $n - 2$ items left in the dearrangement, a value given by g_{n-2}. Summing these disjoint possibilities shows that

$$(2.9) \qquad\qquad g_n = (n - 1)(g_{n-1} + g_{n-2})$$

The number of dearrangements grows quickly with n as shown in the following table:

n	2	4	6	8	10	12	14
g_n	1	9	265	14,833	1,334,961	176,214,841	32,071,101,049

A billion dearrangements are surpassed with $n = 13$ (2,290,792,932 ways to be exact). Thus, if you take a fresh pack of cards, separate out one suit in its numeric order and shuffle these 13 cards thoroughly, then about 36.787% of 13! possibilities will correspond to dearrangements. In fact, it is not too difficult to show that as n gets large, the fraction of random permutations of n items that are dearrangements converges to $1/e$ where $e \approx 2.718281828$ is called *Euler's number* after the Swiss mathematician Leonard Euler (1707–1783) who, among his vast achievements, studied properties of the exponential function.[7]

The difficulty with solving recurrence (2.9) lies in the multiplicative factor $(n - 1)$ found in the equation. To counteract this, consider a scaled version where $f_n = g_n/n!$. This produces the recurrence

$$
\begin{aligned}
f_n = \frac{g_n}{n!} &= \frac{n-1}{n!}\,(g_{n-1} + g_{n-2}) \\
&= \frac{n-1}{n!}\,((n-1)!f_{n-1} + (n-2)!f_{n-2}) \\
&= \left(1 - \frac{1}{n}\right) f_{n-1} + \frac{1}{n} f_{n-2}
\end{aligned}
$$

Rewriting this reveals a difference that can be formed between successive index values

$$f_n - f_{n-1} = -\frac{1}{n}\,(f_{n-1} - f_{n-2})$$

---

[7]Convergence is quick. The value of $|g_{13}/13! - 1/e|$ is about 10^{-11}.

This can be iterated to yield

$$f_n - f_{n-1} = (-1)^n \frac{1}{n!}$$

This telescopes to the boundary $f_0 = 1$ leading to the following summation:

$$f_n = \sum_{i=0}^{n} \frac{(-1)^i}{i!}$$

Converting back to the original recursion produces a lovely result: the number of dearrangements equals an alternating sum of permutations:

$$(2.10) \qquad g_n = n! f_n = \sum_{i=0}^{n} (-1)^i p_{i,n}$$

2.2 Combinations: Without Replacement

Consider a particular combination obtained from selecting k items from a set of size n. The ordering of the items in this combination can be permuted in $k!$ ways without adding to the number of combinations. Thus an equation for the number of possible combinations, denoted by $c_{k,n}$, is given by

$$(2.11) \qquad c_{k,n} = \frac{n^{\underline{k}}}{k!} = \prod_{j=0}^{k-1} \frac{n-j}{k-j}$$

It is customary to write this using a *binomial coefficient*[8]

$$(2.12) \qquad c_{k,n} = \binom{n}{k} = \frac{n!}{(n-k)!\,k!}$$

---

[8] The product expansion (2.11) is used when computing the value of a binomial coefficient.

Values of $c_{k,n}$ for small parameter values are given in the following table:

Values of $c_{k,n} = \dbinom{n}{k}$								
$k \backslash n$	1	2	3	4	5	6	7	8	9
0	1	1	1	1	1	1	1	1	1
1	1	2	3	4	5	6	7	8	9
2		1	3	6	10	15	21	28	36
3			1	4	10	20	35	56	84
4				1	5	15	35	70	126
5					1	6	21	56	126
6						1	7	28	84
7							1	8	36
8								1	9
9									1

A recursive derivation of (2.11) is instructive and proves to be useful in future derivations. Initial conditions are easily calculated: $c_{k,n} = 0$ for $k > n$ or $k < 0$, $c_{1,n} = n$ and $c_{n,n} = 1$. To derive a general recurrence for $c_{k,n}$, partition selections into two disjoint sections. If item j is selected, then the remaining $k-1$ items must be selected from the $n-1$ remaining items, a quantity given by $c_{k-1,n-1}$. If item j is not selected, then it is equivalent to not being in the set, a quantity given by $c_{k,n-1}$. The total number of combinations without replacement is the sum of these two disjoint possibilities and thus equals

$$(2.13) \qquad c_{k,n} = c_{k-1,n-1} + c_{k,n-1}$$

Simple algebra establishes that equation (2.11) (or equation (2.12)) satisfies recursion (2.13). The value of $c_{k,n}$ equals the number of ways to pick k items from a set of n or, equivalently, equals the number of possible subsets of size k that can be formed from a set of n items.

2.2.1   Binomial Identities

Straightforward algebra establishes the following identities between binomial coefficients:

$$(2.14) \qquad \binom{n}{k} = \binom{n}{n-k}$$

$$(2.15) \qquad \binom{n}{k} = \binom{n-1}{k-1} + \binom{n-1}{k}$$

$$(2.16) \qquad \binom{n}{k} = \frac{n}{k}\binom{n-1}{k-1} = \frac{n-k+1}{k}\binom{n}{k-1}$$

$$(2.17) \qquad \binom{n}{k}\binom{k}{j} = \binom{n}{j}\binom{n-j}{k-j}$$

Identity (2.14) shows the symmetry of binomials coefficients and identity (2.15) is a direct restatement of recurrence (2.13). Identity (2.16) follows from the product expansion of (2.11) and from setting $j = 1$ in (2.17). A typical application of (2.17) is to separate variables j and k in a summation. As an example, consider the following derivation which uses (2.17) in the first step:

$$(2.18) \qquad \sum_{k=0}^{n}\binom{n}{k}\frac{a^k}{1+k} = \frac{1}{n+1}\sum_{k=0}^{n}\binom{n+1}{k+1}a^k$$

$$= \frac{1}{n+1}\sum_{j=1}^{n+1}\binom{n+1}{j}a^{j-1}$$

$$= \frac{(1+a)^{n+1}-1}{a(1+n)}$$

Two special cases of this identity arise when $a = 1$ or $a = -1$:

$$(2.19) \qquad \sum_{k=0}^{n}\binom{n}{k}\frac{1}{1+k} = \frac{2^{n+1}-1}{n+1}$$

$$(2.20) \qquad \sum_{k=0}^{n}(-1)^k\binom{n}{k}\frac{1}{1+k} = \frac{1}{n+1}$$

Summations of binomial coefficients are typically derived using induction from an easily calculated base class. There are two ways to interpret the recursion of equation (2.13). The *backward view* starts from $c_{k,n}$ and recurses to a lower value of n with the values $c_{k-1,n-1}$ and $c_{k,n-1}$. This generates an identity which starts from the base case, $c_{0,2} + c_{1,2} + c_{2,2} = 4$, suggesting that

$$(2.21) \qquad \sum_{k=0}^{n}\binom{n}{k} = 2^n$$

Assume this is true for all values less than or equal to n. Using identity (2.15) implies that

$$\sum_{k=0}^{n+1}\binom{n+1}{k} = \sum_{k=0}^{n+1}\binom{n}{k-1} + \binom{n}{k}$$

$$= \sum_{k=0}^{n+1}\binom{n}{k-1} + 2^n$$

$$= \sum_{k=0}^{n}\binom{n}{k} + \binom{n}{-1} + 2^n$$

$$= 2^n + 0 + 2^n$$

$$= 2^{n+1}$$

thus proving the claim.

A variation of identity (2.21) involves summing over even or odd indexed values of k. Notice that small examples include: $c_{0,3} + c_{2,3} = 4$ and $c_{0,4} + c_{2,4} + c_{4,4} = 8$ which suggest that

(2.22)
$$\sum_{k \text{ even}}\binom{n}{k} = 2^{n-1}$$

Also note that $c_{1,3} + c_{3,3} = 4$ and $c_{1,4} + c_{3,4} = 8$ which suggests that

(2.23)
$$\sum_{k \text{ odd}}\binom{n}{k} = 2^{n-1}$$

Assume that both of these assumptions holds for all values up to n and calculate

$$\sum_{k \text{ even}}\binom{n+1}{k} = \sum_{k \text{ even}}\binom{n}{k-1} + \binom{n}{k}$$

$$= 2^{n-1} + \sum_{k \text{ even}}\binom{n}{k-1}$$

$$= 2^{n-1} + \sum_{k \text{ odd}}\binom{n}{k}$$

$$= 2^{n-1} + 2^{n-1} = 2^n$$

A similar argument holds if the initial summation takes place over odd indices. This establishes the double induction and proves identities (2.22) and (2.23).

Equations (2.21), (2.22), and (2.23) are all special cases of a much deeper result—the *binomial theorem*. This states that

$$(2.24) \qquad (x+y)^n = \sum_{k=0}^{n} \binom{n}{k} x^k y^{n-k}$$

This is easily established for small n values. Assume it holds for all values up to n. Then identity (2.15) shows that the pattern continues:

$$(x+y)^{n+1} = \sum_{k=0}^{n+1} \binom{n+1}{k} x^k y^{n+1-k}$$

$$= \sum_{k=0}^{n+1} \left(\binom{n}{k} + \binom{n}{k-1} \right) x^k y^{n+1-k}$$

$$= y \sum_{k=0}^{n} \binom{n}{k} x^k y^{n-k} + x \sum_{k=1}^{n+1} \binom{n}{k-1} x^{k-1} y^{n+1-k}$$

$$= y(x+y)^n + x \sum_{\ell=0}^{n} \binom{n}{\ell} x^k y^{n-k}$$

$$= y(x+y)^n + x(x+y)^n = (x+y)^{n+1}$$

There are a countless number of identities that arise by varying parameters of the binomial theorem besides those just mentioned. Listing just a few:

$$(2.25) \qquad \left(1 - \frac{1}{\ell}\right)^n = \sum_{k=0}^{n} \binom{n}{k} \left(\frac{-1}{\ell}\right)^k$$

$$(2.26) \qquad (x+a)^n - (x-a)^n = 2 \sum_{k \text{ odd}} \binom{n}{k} x^{n-k} a^k$$

$$(2.27) \qquad \left(\frac{1}{x} + \frac{1}{y}\right)^n = \frac{1}{y^n} \sum_{k=0}^{n} \binom{n}{k} \left(\frac{y}{x}\right)^k$$

$$(2.28) \qquad \sum_{k=0}^{n-i}(-1)^k \binom{n-i}{k} = \begin{cases} 0, \ i = 0, \ldots, n-1 \\ 1, \ i = n \end{cases}$$

A further identity can be obtained by forming a telescoping summation combined with the binomial theorem. To derive this, observe that

$$n = \sum_{k=1}^{n}\binom{n}{k}k - \binom{n}{k-1}(k-1)$$

$$= \sum_{k=1}^{n}\binom{n}{k-1}(n-2k+2) \qquad \text{From Identity (2.16)}$$

$$= n + \sum_{k=1}^{n+1}\binom{n}{k-1}(n-2k+2)$$

$$= n + \sum_{k=0}^{n}\binom{n}{k}(n-2k)$$

Thus

$$\sum_{k=0}^{n}\binom{n}{k}(n-2k) = 0$$

which, using the binomial theorem, establishes the identity

$$(2.29) \qquad \sum_{k=0}^{n}\binom{n}{k}k = n2^{n-1}$$

A similar telescoping summation can be used to calculate

$$(2.30) \qquad \sum_{k=0}^{n}\binom{n}{k}k(k-1) = n(n-1)2^{n-2}$$

and

$$(2.31) \qquad \sum_{k=0}^{n}\binom{n}{k}k^2 = n(n+1)2^{n-2}$$

The next identity we derive is the *forward view* of recurrence (2.13) that proceeds from $n-1$ to n and corresponds to a summation of the numerator of the combinatorial coefficient. This summation follows directly from the listing recursions from (2.13):

$$c_{k+1,n+1} = c_{k,n} + c_{k+1,n}$$

$$c_{k+1,n} = c_{k,n-1} + c_{k+1,n-1}$$

$$c_{k+1,n-1} = c_{k,n-2} + c_{k+1,n-2}$$

$$\vdots \qquad \vdots \qquad \vdots$$

$$c_{k+1,k+2} = c_{k,k+1} + c_{k+1,k+1}$$

Collecting these yields

$$c_{k+1,n+1} = c_{k,n} + c_{k,n-1} + \cdots + c_{k,k+1} + 1$$

implies that

(2.32)
$$\binom{n+1}{k+1} = \sum_{\ell=k}^{n} \binom{\ell}{k}$$

With the index substitution, $j = \ell - k$, the right-hand side of identity (2.32) can be rewritten as

(2.33)
$$\sum_{\ell=k}^{n} \binom{\ell}{k} = \sum_{j=0}^{n-k} \binom{k+j}{k}$$

Making the substitution $m - 1 = n - k$, with equations (2.32) and (2.33), shows that[9]

(2.34)
$$\sum_{r=0}^{m-1} \binom{k+r}{k} = \binom{m+k}{k+1}$$

Combinatoric arguments can often provide insight into identities without the need for algebraic manipulation. Consider selecting k items

---

[9]This equation is termed Pascal's equation after the French mathematician Blaise Pascal (1623–1662).

from a list of n. Divide the list n into two sublists of size m and $n - m$ where $0 \leq m \leq n$ and suppose that j items are selected from the first list and $k - j$ items from the second list. There are $c_{j,m} c_{k-j,n-m}$ ways in which such a selection can be made. Accounting for all such possibilities leads to the identity

$$(2.35) \qquad \binom{n}{k} = \sum_{j=0}^{n} \binom{m}{j} \binom{n-m}{k-j}, \quad 0 \leq m \leq n$$

The conditions we have placed on binomial terms so that $c_{k,n} = 0$ if $k < 0$, $n < 0$ or $k > n$ allow us to write the right-hand side of equation (2.35) without needing to put restrictions on the binomial coefficients contained in the summation. This identity is termed *Vandermonde convolution* named after Alexandre-Théophile Vandermonde (1735–1796), a French mathematician who supported the French revolution of 1789 and is most known for his mathematical work in determinants.

2.3 Combinations with Replacement

This leaves the last problem—to calculate the number of combinations obtained when one uses a replacement strategy. To derive an equation, let $c_{k,n}^r$ denote the number of combinations obtained when selecting k items from a set of n items using replacement.[10] We set $c_{0,1}^r = 1$ and it is clear that $c_{1,n}^r = n$ and $c_{k,1}^r = 1$. To derive a general equation for $c_{k,n}^r$, partition selections into two disjoint sections. If item j is selected, then the number of combinations with replacement is the same as if it had been selected on the first selection, a quantity given by $c_{k-1,n}^r$. On the other hand, if item j is not selected, then it is equivalent to not being in the set, a quantity given by $c_{k,n-1}^r$. The total number of combinations is the sum of these two disjoint possibilities and thus equals

$$(2.36) \qquad\qquad c_{k,n}^r = c_{k-1,n}^r + c_{k,n-1}^r$$

---

[10]Again, the r superscript means *replacement* rather than being an integer index.

It is interesting to compare equations (2.13) and (2.36). They differ only in the indices of the first term. The n index of $c_{k-1,n}^r$ as compared to the $n-1$ index of $c_{k-1,n-1}$ is a result of putting a selected item back in the list when the replacement policy is utilized. This results in a substantial increase in the number of possibilities that a replacement policy has in comparison to a non-replacement policy.

To proceed with the derivation, consider equation (2.36) where $k=2$:

$$c_{2,n}^r = c_{1,n}^r + c_{2,n-1}^r$$
$$= c_{1,n}^r + c_{1,n-1}^r + c_{2,n-2}^r$$

Observe that $c_{2,n-2}^r$ is of the same form as $c_{2,n}^r$ except it moves 2 down on the n value which suggests iterating

$$c_{2,n}^r = c_{1,n}^r + c_{1,n-1}^r + \cdots + c_{1,2}^r + c_{1,1}^r$$
$$= n + (n-1) + \cdots + 2 + 1$$
$$= \binom{n+1}{2}$$

This case suggests the following guess for a general solution:

$$c_{k,n}^r = \binom{n+k-1}{k}$$

Assume this equation holds for all values less than or equal to some value k. Then, using the recurrence (2.36) permits

$$c_{k+1,n}^r = c_{k,n}^r + c_{k+1,n-1}^r$$
$$= c_{k,n}^r + c_{k,n-1}^r + c_{k+1,n-2}^r$$
$$= c_{k,n}^r + c_{k,n-1}^r \cdots + c_{k,2}^r + c_{k+1,1}^r$$
$$= \binom{k+n-1}{k} + \cdots + \binom{k+2}{k} + \binom{k+1}{k} + 1$$
$$= \sum_{r=0}^{n-1} \binom{k+r}{k} = \binom{n+(k+1)-1}{k+1}$$

where the last equation uses the identity (2.34). This shows that the induction is satisfied and thus that

$$(2.37) \qquad c_{k,n}^r = \left\langle \begin{matrix} n \\ k \end{matrix} \right\rangle$$

where *binomial-R coefficients* are defined by

$$(2.38) \qquad \left\langle \begin{matrix} n \\ k \end{matrix} \right\rangle = \binom{n+k-1}{k}$$

Values of $c_{k,n}^r$ for small parameter values are given in the following table:

	Values of $c_{k,n}^r = \left\langle \begin{matrix} n \\ k \end{matrix} \right\rangle$							
$k \backslash n$	1	2	3	4	5	6	7	8	9
0	1	1	1	1	1	1	1	1	1
1	1	2	3	4	5	6	7	8	9
2	1	3	6	10	15	21	28	36	45
3	1	4	10	20	35	56	84	120	165
4	1	5	15	35	70	126	210	330	495
5	1	6	21	56	126	252	462	792	1,287
6	1	7	28	84	210	462	924	1,716	3,003
7	1	8	36	120	330	792	1,716	3,432	6,435
8	1	9	45	165	495	1,287	3,003	6,435	12,870
9	1	10	55	220	715	2,002	5,005	11,440	24,310

2.3.1 Binomial-R Identities

Some identities that are easy to verify include

$$(2.39) \qquad \left\langle \begin{matrix} n \\ k \end{matrix} \right\rangle = \left\langle \begin{matrix} k+1 \\ n-1 \end{matrix} \right\rangle$$

$$\left\langle \begin{matrix} n \\ k \end{matrix} \right\rangle = \left\langle \begin{matrix} n \\ k-1 \end{matrix} \right\rangle + \left\langle \begin{matrix} n-1 \\ k \end{matrix} \right\rangle$$

$$\left\langle\begin{matrix}n\\k\end{matrix}\right\rangle = \frac{n}{k}\left\langle\begin{matrix}n+1\\k-1\end{matrix}\right\rangle$$

$$\binom{n}{k} = \left\langle\begin{matrix}n+1-k\\k\end{matrix}\right\rangle$$

Adding one more term to (2.34) produces a similar identity expressed in terms of binomial-R coefficients:

$$(2.40)\qquad \sum_{r=0}^{n}\binom{k+r}{k} = \sum_{r=0}^{n-1}\binom{k+r}{k} + \binom{n+k}{k}$$

$$= \binom{n+k}{k+1} + \binom{n+k}{k}$$

$$= \binom{n+k+1}{k+1}$$

$$= \left\langle\begin{matrix}n+1\\k+1\end{matrix}\right\rangle$$

The flow of this derivation uses equations from (2.16) and (2.34). One can also derive an identity involving the sum binomial-R coefficients given by

$$(2.41)\qquad \sum_{k=0}^{m}\left\langle\begin{matrix}n\\k\end{matrix}\right\rangle = \left\langle\begin{matrix}n+1\\m\end{matrix}\right\rangle$$

The size difference between combinations without and with replacement can be quantified, similar to that of equation (2.8), by forming their ratio:

$$(2.42)\qquad \frac{c_{k,n}}{c_{k,n}^{r}} = \prod_{j=1}^{k-1}\left(1 - \frac{k}{n+k-j}\right)$$

With the same values as in the birthday problem, $k = 23$ and $n = 365$, the ratio of combinations given in equation (2.42) yields an answer of around 25% in comparison to 50% found with permutations.

2.3.2 Polynomial Solutions to Combinatorial Problems

To view the previous analysis within a general framework, suppose there are a set of n integers, m_i, $i = 1, \ldots, n$, each between 0 and k having a total sum that equals k:

$$m_1 + m_2 + \cdots + m_n = k, \qquad 0 \leq m_i \leq k$$

How many ways can such numbers be selected to satisfy these constraints? This problem can be thought as the number of n partitions of the integer k. For example, if $n = 3$ and $k = 2$, then there are six possible partitions:

(2.43) $\{2,0,0\}$, $\{1,1,0\}$, $\{1,0,1\}$, $\{0,2,0\}$, $\{0,1,1\}$, $\{0,0,2\}$

To answer the general question, suppose that k balls are thrown randomly into n buckets. The value of m_i counts the total number of balls that land in bucket i. Some thought shows that this is equivalent to the number of combinations for selecting k items from a set of n where a replacement strategy is used. To see this, associate a bucket with each item in the set of n items. With this association, selecting the i'th item in the set is equivalent to throwing a ball into the i'th bucket. The constraints on the buckets mean that only up to k balls can land in any particular bucket. Thus the solution to the posed question is that the number of possible partitions equals $c_{k,n}^r$. This simple solution leads to an extremely useful concept which will be derived in the following paragraphs.

The great thing about being a mathematician is that your work, which is really like play, can be done almost anywhere so there is never a danger of becoming bored. Perhaps, for instance, you are stuck in the middle of a theater during a particularly uninteresting play. Then, as long as you have a pen, you can play with equations on the back or margins of the program. You might, during one of these occasions, jot down a simple infinite polynomial like

$$f(x) = 1 + x + x^2 + x^3 + \cdots$$

If x is between 0 and 1, then such a sequence convergences and its sum is given by

$$\frac{1}{1-x} = \sum_{i=0}^{\infty} x^i$$

The linkage between this and combinatorics arises when f is raised to a power

$$f^n(x) = \left(\frac{1}{1-x}\right)^n$$

It is clear that this expression is another infinite polynomial

$$f^n(x) = \sum_{i=0}^{\infty} a_{i,n} x^i$$

with integer coefficients $a_{i,n}$. What are these coefficients?

As a concrete example, suppose that $n = 3$ and $k = 2$. Consider the value of $a_{2,3}$ which corresponds to the coefficient of x^2 in $f^3(x)$. To facilitate the argument, write

$$f^3(x)$$
$$= \underbrace{(x^0 + x^1 + x^2 + \cdots)}_{\text{first group}} \times \underbrace{(x^0 + x^1 + x^2 + \cdots)}_{\text{second group}} \times \underbrace{(x^0 + x^1 + x^2 + \cdots)}_{\text{third group}}$$

Let the exponent of x selected in the first group be denoted by m_1 and similarly define m_2 and m_3. When the sum of these x exponents, $m_1 + m_2 + m_3$, equals 2 then those factors contribute to the value of $a_{2,3}$. Since the coefficients of $f(x)$ are all 1, the value of $a_{2,3}$ is the number of ways three non-negative integers sum to 2. In our example, this corresponds to the sets given in (2.43). The solution to the problem posed a couple of paragraphs back showed that this quantity equals $c_{2,3}^r$. Clearly this argument generalizes, and thus $a_{i,n} = c_{i,n}^r$ which establishes the equation:

$$(2.44) \qquad f^n(x) = \sum_{i=0}^{\infty} \left\langle \begin{array}{c} n \\ i \end{array} \right\rangle x^i$$

To link this analysis back to counting, recall that $c_{k,n}^r$ is the number of ways that $m_1 + m_2 + \cdots + m_n = k$ where m_i was bounded below by 0 and above by k. Suppose that the value of m_i has a different set of constraints. For example, consider a simple case where $n = 3$ and $k = 2$ that has the following constraints:

$$m_1 \in \{0, 1\}, \qquad m_2 \in \{1, 2\}, \qquad m_3 \in \{0, 2\}$$

The number of combinations where $m_1 + m_2 + m_3 = 2$ can be easily listed

(2.45) $\{1,1,0\}, \quad \{0,2,0\}$

But is there a way to calculate that there are 2 possibilities without actually listing them all?

The answer lies in the previous analysis. Associate the polynomial $x^0 + x^1$ with the first value of m_1 which mathematically incorporates the constraint that $m_1 \in \{0,1\}$. Similarly associate the polynomial $x^1 + x^2$ with m_2 and $x^0 + x^2$ with m_3. The polynomial $g(x)$ defined by the product of all these sub polynomials is given by

(2.46) $g(x) = (x^0 + x^1)(x^1 + x^2)(x^0 + x^2) = x + 2x^2 + 2x^3 + 2x^4 + x^5$

By construction, the coefficient of x^ℓ corresponds to the number of possible combinations that result when $m_1 + \ldots + m_k = \ell$. Hence, the expansion in equation (2.46) shows that the number of such possibilities equals 2 when $\ell = 2$, thus confirming the enumeration given in (2.45). Additionally, the expansion shows that there are no ways to sum to 0, one way to sum to 1 or 5, and two ways to sum to 2, 3, or 4.

To state the general case, suppose that m_i can only have values

(2.47) $m_i \in \{v_{i,1}, v_{i,2}, \ldots, v_{i,n_i}\}$

and define

$$g_i(x) = x^{v_{i,1}} + x^{v_{i,2}} + \cdots + x^{v_{i,n_i}}$$

from which the following product is formed:

$$g(x) = \prod_{i=1}^{n} g_i(x)$$

Then ,the coefficient of x^ℓ in the polynomial $g(x)$ equals the number of ways that $m_1 + m_2 + \cdots + m_n = \ell$ where each m_i satisfies its particular set of constraints (2.47). In essence, this technique solves an entire class of difficult problems—a strikingly deep result considering that it arises from a simple combinatoric derivation.

As a final example, consider the number of different values that can be obtained by summing combinations of the first n primes, $p_i, i = 1, \ldots, n$. Each prime can either be selected in the sum or not and thus $g_i(x) = 1 + x^{p_i}$, $i = 1, \ldots, n$, and

$$g(x) = \prod_{i=1}^{n} 1 + x^{p_i}$$

For the special case of $n = 4$ this expansion yields

$$g(x) = 1 + x^2 + x^3 + 2x^5 + 2x^7 + x^8 + x^9 + 2x^{10} + 2x^{12} + x^{14} + x^{15} + x^{17}$$

There are multiple results contained in this expression: there are two ways to sum the first four primes (2,3,5,7) leading to integers in set $\{5, 7, 10, 12\}$ and one way for integers in set $\{0, 2, 3, 8, 9, 14, 15, 17\}$. The set of integers $\{4, 6, 11, 13, 16, 18, 19, \ldots\}$ has been left out of the set of possibilities and there are 12 different sums that are possible since this equals the number of terms contained in $g(x)$. The number of prime numbers contained in the set of possibilities equals 5 (the set $\{2, 3, 5, 7, 17\}$) and the largest consecutive sequence of numbers has four members (the series 7, 8, 9, 10). Clearly, these observations open new questions concerning how they scale as n increases. Is the longest consecutive series, for instance, bounded if n increases indefinitely?

2.4 Transforms and Identities

Let $x = (x_0, \ldots, x_n)$ be a vector of length $n + 1$ and define functions $f_\ell(x)$ and $g_\ell(x)$ as follows:

(2.48) $$f_\ell(x) = \sum_{k=0}^{\ell} \binom{\ell}{k} x_k$$

(2.49) $$g_\ell(x) = \sum_{k=0}^{\ell} (-1)^{\ell-k} \binom{\ell}{k} x_k, \qquad \ell = 0, \ldots, n$$

Define $a = (a_0, \ldots, a_n)$ and $b = (b_0, \ldots, b_n)$ and set $b_\ell = f(a)$, $\ell = 0, \ldots, n$. Expanding (2.49) yields

$$g_\ell(\boldsymbol{b}) = \sum_{k=0}^{\ell} (-1)^{\ell-k} \binom{\ell}{k} b_k$$

$$= \sum_{k=0}^{\ell} (-1)^{\ell-k} \binom{\ell}{k} \sum_{i=0}^{k} \binom{k}{i} a_i$$

$$= \sum_{i=0}^{\ell} a_i \binom{\ell}{i} \sum_{k=i}^{\ell} \binom{\ell-i}{k-i} (-1)^{\ell-k}, \quad \text{From identity (2.17)}$$

$$= \sum_{i=0}^{\ell} a_i \binom{\ell}{i} \sum_{j=0}^{\ell-i} \binom{\ell-i}{j} (-1)^{\ell-j-i}$$

$$= \sum_{i=0}^{\ell} a_i (-1)^{\ell-i} \binom{\ell}{i} \sum_{j=0}^{\ell-i} \binom{\ell-i}{j} (-1)^{j}$$

$$= a_\ell \qquad\qquad\qquad\qquad\qquad\qquad \text{From identity (2.28)}$$

The paired equations $b_\ell = f_\ell(\boldsymbol{a})$ and $a_\ell = g_\ell(\boldsymbol{b})$ show that functions f_ℓ and g_ℓ are inverse functions of each other. This observation creates a useful transformation termed *binomial transformation*. To distinguish these functions, f_ℓ is typically termed a *binomial transform* and g_ℓ a *binomial inverse*.

Binomial transformation can be used to create a paired set of identities whenever a sequence satisfies either (2.48) or (2.49). For example, equation (2.31) corresponds to the binomial transform of $a_\ell = \ell^2$. This implies that $b_\ell = \ell(\ell+1)2^{\ell-2}$, $\ell = 0, \ldots, n$ thus creating the paired identity

$$(2.50) \qquad n^2 = \sum_{k=0}^{n} \binom{n}{k} (-1)^{n-k} k(k+1)2^{k-2}$$

The inverse transforms of (2.21), (2.25), (2.29), and (2.30) are given by

$$(2.51) \qquad \sum_{k=0}^{n} \binom{n}{k} (-1)^{n-k} 2^k = 1$$

$$(2.52) \qquad \sum_{k=0}^{n} \binom{n}{k} (-1)^{n-k} \left(1 - \frac{1}{\ell}\right)^k = \left(\frac{-1}{\ell}\right)^n$$

$$\text{(2.53)} \qquad \sum_{k=0}^{n} \binom{n}{k} (-1)^{n-k} k 2^{k-1} = n$$

$$\text{(2.54)} \qquad \sum_{k=0}^{n} \binom{n}{k} (-1)^{n-k} k(k-1) 2^{k-2} = n(n-1)$$

The inverse transform of (2.18) provides the paired identity

$$\text{(2.55)} \qquad \sum_{k=0}^{n} \binom{n}{k} (-1)^{n-k} \frac{(1+a)^{k+1} - 1}{k+1} = \frac{a^{n+1}}{n+1}$$

A special case of (2.18) and (2.55) for $a = 2$ creates the following two identities:

$$\text{(2.56)} \qquad \sum_{k=0}^{n} \binom{n}{k} \frac{2^{k+1}}{k+1} = \frac{3^{n+1} - 1}{n+1}$$

$$\text{(2.57)} \qquad \sum_{k=0}^{n} \binom{n}{k} (-1)^{n-k} \frac{3^{k+1} - 1}{k+1} = \frac{2^{n+1}}{n+1}$$

Another form of a binomial transform is given by

$$\text{(2.58)} \qquad h_\ell(\boldsymbol{x}) = \sum_{k=0}^{\ell} (-1)^k \binom{\ell}{k} x_k$$

which defines an *involution*, a function that is its own inverse. To establish this, let $b_\ell = h_\ell(\boldsymbol{a})$ and proceed as follows:

$$h_\ell(\boldsymbol{b}) = \sum_{k=0}^{\ell} (-1)^k \binom{\ell}{k} b_k$$

$$= \sum_{k=0}^{\ell} (-1)^k \binom{\ell}{k} \sum_{i=0}^{k} (-1)^i \binom{k}{i} a_i$$

$$= \sum_{i=0}^{\ell} a_i \binom{\ell}{i} \sum_{k=i}^{\ell} \binom{\ell - i}{k - i} (-1)^{k+i}$$

$$= \sum_{i=0}^{\ell} a_i \binom{\ell}{i} \sum_{j=0}^{\ell-i} \binom{\ell-i}{j} (-1)^j$$

$$= a_\ell$$

To illustrate a use of this form of binomial transformation, substitute $a = -b$ in (2.18) which then implies the two identities

(2.59) $$\sum_{k=0}^{n} (-1)^k \binom{n}{k} \frac{b^k}{k+1} = -\frac{(1-b)^{n+1} - 1}{b(n+1)}$$

and

(2.60) $$\sum_{k=0}^{n} (-1)^k \binom{n}{k} \frac{(1-b)^{k+1} - 1}{b(k+1)} = -\frac{b^n}{(n+1)}$$

As a special case of these identities, set $b = 2$ which then creates identities

(2.61) $$\sum_{k=0}^{n} (-1)^k \binom{n}{k} \frac{2^k}{k+1} = \begin{cases} 0, & n \text{ odd}, \\ \frac{1}{n+1}, & n \text{ even} \end{cases}$$

(2.62) $$\sum_{k=0, \; k \text{ even}}^{n} (-1)^k \binom{n}{k} \frac{1}{(k+1)} = \frac{2^n}{n+1}$$

Chapter 3
Syntax Precedes Semantics

"Good Morning!" said Bilbo, and he meant it. The sun was shining,and the grass was very green. But Gandalf looked at him from under long bushy eyebrows that stuck out further than the brim of his shady hat.

"What do you mean?" he said. "Do you wish me a good morning,or mean that it is a good morning whether I want it or not;or that you feel good this morning; or that it is a morning to be good on?"

"All of them at once," said Bilbo.

J. R. R. Tolkien, The Hobbit

How you say something is often as important as what you say. A simple *"Good Morning"* can confuse even a wizard like Gandalf and this can be no more apparent than in writing mathematics where ambiguity is not tolerated. This explains one reason why LaTeX has made such a major impact on mathematics even though it only deals with the syntax of mathematical writing and not its content. The TeX project started by Donald Knuth (1938–) gave mathematicians the tools they needed to be able to write beautifully typeset papers and books that brought to light the semantics of math in a crystal clear format. In this way, syntax precedes semantics.

Notation is also a vitally important component of mathematics. Clear notation reveals patterns to the mind that are obscured by more awkward expressions. To illustrate this, recall that Pascal's equation was derived in the chapter, *Let Me Count the Ways* with equation (2.34). To express this more concisely recall that the falling factorial notation is defined by

$$(3.1) \qquad n^{\underline{k}} = n(n-1)\cdots(n-k+1)$$

© Springer Nature Switzerland AG 2020
R. Nelson, *A Brief Journey in Discrete Mathematics*,
https://doi.org/10.1007/978-3-030-37861-5_3

and the rising factorial notation by

$$(3.2) \qquad n^{\overline{k}} = n(n+1)\cdots(n+k-1)$$

Some algebra shows that $n^{\overline{k}} = (n+k-1)^{\underline{k}}$ and $n^{\underline{k}} = (n-k+1)^{\overline{k}}$. With these notations we can write a binomial coefficient in multiple ways

$$\binom{n}{k} = \frac{n!}{k!(n-k)!} = \frac{n^{\underline{k}}}{k!} = \frac{(n-k+1)^{\overline{k}}}{k!}$$

The combinatorial term on the left-hand side of (2.34) can now be expressed as

$$\binom{i+k-1}{i-1} = \frac{i^{\overline{k}}}{k!}$$

and the right-hand side by

$$\binom{n+k}{k+1} = \binom{n-1+k+1}{k+1} = \frac{n^{\overline{k+1}}}{(k+1)!}$$

Some minor simplifications then shows that Pascal's formula (2.34) can be expressed compactly as

$$(3.3) \qquad \sum_{i=1}^{n} i^{\overline{k}} = \frac{n^{\overline{k+1}}}{k+1}$$

Equation (3.3) expresses Pascal's equation in a form that highlights a pattern which is not evident in (2.34) and expresses a relationship contained in the integers. Define a *variety k* integer to be the product of k successive integers. Thus $i^{\overline{k}}$ is the i'th *variety k* integer. In these terms, equation (3.3) expresses a relationship between variety k integers with variety $k+1$ integers. Specifically, the equation shows that the sum of the first n *variety k* integers equals the $n+1$'st variety $k+1$ integer divided by $k+1$. This result will be partially generalized later in the book with equation (7.26) which considers the powers of *variety* 2 integers. The next section brings up the question: what is a rising factorial?

3.1 Stirling Numbers of the First Kind

To move towards answering this, first note that $n^{\overline{k}}$ is a k'th degree polynomial in n. Let the coefficients of this polynomial be denoted by $b_{i,k}$ for $i = 0, \ldots k$. Two coefficients are immediately obvious: $b_{0,k} = 0$ and $b_{k,k} = 1$. What are the remaining coefficients? To answer this, a straightforward calculation shows that

	Coefficients $b_{i,k}$					
$k \backslash i$	1	2	3	4	5	Sum
1	1					1
2	1	1				2
3	2	3	1			6
4	6	11	6	1		24
5	24	50	35	10	1	120

Blanks above equal 0 and thus, as an illustration, the table shows that $n^{\overline{4}} = 6n + 11n^2 + 6n^3 + n^4$.

The numbers in the table have some interesting special values. For example, the sum of the rows (the last column in the table) equals factorial numbers, $\sum_{i=0}^{n} b_{i,n} = n!$. Also, the first column is simply a shifted version of the summation column, $b_{1,n} = (n-1)!$ and the submajor diagonal values in the table correspond to binomial coefficients, $b_{n-1,n} = \binom{n}{2}$. Other values found in the table do not have obvious values which leads us to the problem of finding a relationship between them.

To derive this relationship, note that

$$n^{\overline{k+1}} = (n+k)n^{\overline{k}} = \underbrace{n \cdot n^{\overline{k}}}_{\text{first part}} + \underbrace{k \cdot n^{\overline{k}}}_{\text{second part}}$$

This shows that the coefficient $b_{i,k+1}$ consists of two parts determined by the exponent of n. The first part corresponds to the coefficient of

n^{i-1} in $n^{\overline{k}}$ since $n \cdot n^{i-1} = n^i$, thus yielding a summand of $b_{i-1,k}$. The second part corresponds to multiplying the coefficient of n^i by k yielding a second summand of $k \cdot b_{i,k}$. Combining both summands shows that

$$(3.4) \qquad b_{i,k+1} = \begin{cases} b_{i-1,k} + k b_{i,k}, & i = 1, \ldots, k \\ \\ 1, & i = k+1 \end{cases}$$

These coefficients frequently appear in mathematics and are termed *Stirling numbers of the first kind*, named after the mathematician James Stirling (1692–1770). This brings up the immediate question: *What are Stirling numbers of the second kind?* We will get to that in a moment.

The typical notation for Stirling numbers of the first kind replaces the parenthesis of binomial coefficients with brackets leading to

$$b_{i,k} = \begin{bmatrix} k \\ i \end{bmatrix}$$

In this notation, the special cases previously mentioned are written as (3.5)

$$\begin{bmatrix} k \\ 0 \end{bmatrix} = 0, \quad \begin{bmatrix} k \\ k \end{bmatrix} = 1, \quad \begin{bmatrix} k \\ 1 \end{bmatrix} = (k-1)!, \quad \begin{bmatrix} k \\ k-1 \end{bmatrix} = \binom{k}{2}, \quad \sum_{i=0}^{k} \begin{bmatrix} k \\ i \end{bmatrix} = k!$$

The recurrence relationship (3.4), implies that

$$(3.6) \qquad \begin{bmatrix} k+1 \\ i \end{bmatrix} = \begin{bmatrix} k \\ i-1 \end{bmatrix} + k \begin{bmatrix} k \\ i \end{bmatrix}$$

and

$$(3.7) \qquad n^{\overline{k}} = \sum_{i=1}^{k} \begin{bmatrix} k \\ i \end{bmatrix} n^i$$

Programming the relationship (3.6), along with the special cases just mentioned, yields the following table:

				Table of $\begin{bmatrix} k \\ i \end{bmatrix}$				
$k \backslash i$	1	2	3	4	5	6	7	8
1	1							
2	1	1						
3	2	3	1					
4	6	11	6	1				
5	24	50	35	10	1			
6	120	274	225	85	15	1		
7	720	1,764	1,624	735	175	21	1	
8	5,040	13,068	13,132	6,769	1,960	322	28	1

Heading back to the modified version of Pascal's equation allows a rewrite in terms of Stirling numbers. The left-hand side of (3.3) can be written as

$$(3.8) \qquad \sum_{i=1}^{n} i^{\overline{k}} = \sum_{i=1}^{n} \sum_{j=1}^{k} \begin{bmatrix} k \\ j \end{bmatrix} i^{j}$$

$$= \sum_{j=1}^{k} \begin{bmatrix} k \\ j \end{bmatrix} \sum_{i=1}^{n} i^{j}$$

$$= \sum_{j=1}^{k} \begin{bmatrix} k \\ j \end{bmatrix} S_{j,n}$$

where

$$S_{j,n} = \sum_{i=1}^{n} i^{j}$$

The right-hand side of the Pascal equation (3.3) implies that

$$(3.9) \qquad \frac{n^{\overline{k+1}}}{k+1} = \frac{1}{k+1} \sum_{j=1}^{k+1} \begin{bmatrix} k+1 \\ j \end{bmatrix} n^{j}$$

Equations (3.3), (3.8), and (3.9) thus create the identity

$$(3.10) \qquad \sum_{j=1}^{k} \begin{bmatrix} k \\ j \end{bmatrix} S_{j,n} = \frac{1}{k+1} \sum_{j=1}^{k+1} \begin{bmatrix} k+1 \\ j \end{bmatrix} n^j$$

Minor modifications to less awkward indices with equations (3.9) and (3.10) show that we have just derived the identity:

$$(3.11) \qquad n^{\overline{k}} = k \sum_{j=1}^{k-1} \begin{bmatrix} k-1 \\ j \end{bmatrix} S_{j,n}$$

and

$$(3.12) \qquad \binom{n+k}{k+1} = \left\langle \begin{matrix} n \\ k+1 \end{matrix} \right\rangle = \frac{1}{k!} \sum_{j=1}^{k} \begin{bmatrix} k \\ j \end{bmatrix} S_{j,n}, \quad k = 1, \dots, n$$

Stirling numbers of the first kind also allow writing falling factorials after minor sign changes. The modified version of (3.7) for falling factorials is given by

$$(3.13) \qquad n^{\underline{k}} = \sum_{i=1}^{k} (-1)^{k-i} \begin{bmatrix} k \\ i \end{bmatrix} n^i$$

As an example, this implies that

$$(3.14) \qquad n^{\underline{4}} = n^4 - 6n^3 + 11n^2 - 6n$$

3.2   Stirling Numbers of the Second Kind

To reverse direction, we seek to derive an equation that expresses a power in terms of falling factorials, specifically

$$(3.15) \qquad n^k = \sum_{i=0}^{k} c_{i,k} n^{\underline{i}}$$

for some unknown constants $c_{i,k}$. Some coefficients are immediately obvious: the boundary cases $c_{0,k} = 0$ and $c_{k,n} = 0, k > n$, which can now be eliminated, and $c_{k,k} = 1$. What are the remaining coefficients? To answer this, a straightforward calculation shows that

	Coefficients $c_{i,k}$				
$k \backslash i$	1	2	3	4	5
1	1				
2	1	1			
3	1	3	1		
4	1	7	6	1	
5	1	15	25	10	1

Blanks above equal 0 and thus, as an illustration, the table shows that $n^4 = n^1 + 7n^2 + 6n^3 + n^4$. There is a repeating pattern to the numbers in this table which can be illustrated by an example. Consider the entry for $c_{3,5} = 25$. It can be written in terms of entries on the preceding row, specifically it equals $3c_{3,4} + c_{3,3} = 3 \times 6 + 7$. This pattern persists in the table which suggests that

$$(3.16) \qquad\qquad c_{i,k} = i c_{i,k-1} + c_{i-1,k-1}$$

Assume that this holds up to some value k and write

$$n^{k+1} = n \, n^k = \sum_{i=1}^{k} c_{i,k} n \, n^i$$

$$= \sum_{i=1}^{k} c_{i,k} \left(n^{i+1} + i \, n^i \right)$$

$$= \sum_{i=1}^{k} c_{i,k} n^{i+1} + \sum_{i=1}^{k} c_{i,k} i n^i$$

$$= \sum_{i=2}^{k+1} c_{i-1,k} n^i + \sum_{i=1}^{k} c_{i,k} i n^i$$

$$= \sum_{i=1}^{k+1} c_{i-1,k} n^i + \sum_{i=1}^{k+1} c_{i,k} i n^i$$

$$= \sum_{i=1}^{k+1} c_{i,k+1} n^i$$

where the last step follows from the induction hypothesis (3.16) and the second to last step is a result of the 0 boundary cases mentioned above.

The coefficients just derived are termed *Stirling numbers of the second kind* and are expressed in combinatorial notation using braces instead of parenthesis, that is $\left\{ {i \atop k} \right\} = c_{i,k}$. This implies that

(3.17)
$$n^k = \sum_{i=1}^{k} \left\{ {k \atop i} \right\} n^{\underline{i}}$$

and, with equation (3.16), that

(3.18)
$$\left\{ {k \atop i} \right\} = i \left\{ {k-1 \atop i} \right\} + \left\{ {k-1 \atop i-1} \right\}$$

Programming the relationship (3.18) along with the special cases just mentioned yields the following table:

Table of $\left\{ {k \atop i} \right\}$							
$k\backslash i$	1	2	3	4	5	6	7	8
1	1							
2	1	1						
3	1	3	1					
4	1	7	6	1				
5	1	15	25	10	1			
6	1	31	90	65	15	1		
7	1	63	301	350	140	21	1	
8	1	127	966	1,701	1,050	266	28	1

Substituting (3.13) into (3.17) implies that

$$n^k = \sum_{i=1}^{k} \left\{ {k \atop i} \right\} n^{\underline{i}}$$

$$= \sum_{i=1}^{k} \left\{ {k \atop i} \right\} \sum_{j=1}^{i} (-1)^{i-j} \left[{i \atop j} \right] n^j$$

$$= \sum_{j=1}^{k} \sum_{i=j}^{k} \left\{ {k \atop i} \right\} (-1)^{i-j} \left[{i \atop j} \right] n^j$$

$$= n^k + \sum_{j=1}^{k-1} n^j \sum_{i=j}^{k} (-1)^{i-j} \left\{ {k \atop i} \right\} \left[{i \atop j} \right]$$

and thus that

(3.19)
$$\sum_{j=1}^{k-1} n^j \sum_{i=j}^{k} (-1)^{i-j} \left\{ {k \atop i} \right\} \left[{i \atop j} \right] = 0$$

Matching powers of n in equation (3.19) produces an equation linking the two kinds of Stirling numbers:

(3.20)
$$\sum_{i=j}^{k} (-1)^{i-j} \left\{ {k \atop i} \right\} \left[{i \atop j} \right] = \begin{cases} 0, & j = 1, \ldots, k-1 \\ 1, & j = k \end{cases}$$

The second linkage between Stirling numbers is seen by comparing the submajor diagonals in tables found on pages 31 and 34 which suggests that

(3.21)
$$\left[{k \atop k-1} \right] = \left\{ {k \atop k-1} \right\}$$

3.2.1 The Stirling Transform and Inverse

Equation (3.20) exposes another example of a transform. To define this let

(3.22)
$$u_\ell(x) = \sum_{k=0}^{\ell} \left\{ {\ell \atop k} \right\} x_k$$

and

(3.23)
$$v_\ell(x) = \sum_{k=0}^{\ell} (-1)^{\ell-k} \left[{\ell \atop k} \right] x_k$$

and set $b_\ell = u_\ell(\boldsymbol{a})$. Similar to the calculation of a binomial transform and its inverse, calculate

$$v_\ell(\boldsymbol{b}) = \sum_{k=0}^{\ell} (-1)^{\ell-k} \begin{bmatrix} \ell \\ k \end{bmatrix} b_k$$

$$= \sum_{k=0}^{\ell} (-1)^{\ell-k} \begin{bmatrix} \ell \\ k \end{bmatrix} \sum_{i=0}^{k} \begin{Bmatrix} k \\ i \end{Bmatrix} a_i$$

$$= \sum_{i=0}^{\ell} a_i \sum_{k=i}^{\ell} (-1)^{\ell-k} \begin{bmatrix} \ell \\ k \end{bmatrix} \begin{Bmatrix} k \\ i \end{Bmatrix}$$

$$= a_\ell \qquad\qquad \text{From Equation (3.20)}$$

This shows that u_ℓ and v_ℓ are inverse function of each other. These functions are termed a Stirling transform and inverse, respectively.

3.3 Combinatorial Interpretation

So far the discussion of Stirling numbers has focused on their algebraic properties. This is manifested in the recurrence relations given by equations (3.6) and (3.18). Like binomial and binomial-R coefficients, however, there is a combinatorial interpretation of these recurrences which lends insight into their associated algebraic properties. Consider, for example, the total number of ways to partition n items into k non-empty sets, a quantity we will denote by $h(k, n)$. To illustrate, for $k = 3$ and set $\{1, 2, 3, 4\}$ the possible partitions are

(3.24) $\{\{1\}, \{2\}, \{3, 4\}\}$ $\{\{1\}, \{3\}, \{2, 4\}\}$ $\{\{1\}, \{4\}, \{2, 3\}\}$
         $\{\{1, 2\}, \{3\}, \{4\}\}$ $\{\{1, 3\}, \{2\}, \{4\}\}$ $\{\{1, 4\}, \{2\}, \{4\}\}$

showing that $h(3, 4) = 6$. With $n = 3$ and set $\{2, 3, 4\}$ the number of partitions with two sets is given by

(3.25) $\{\{2\}, \{3, 4\}\}$ $\{\{3\}, \{2, 4\}\}$ $\{\{4\}, \{2, 3\}\}$

showing that $h(2,3) = 3$ and the number of partitions with three sets by

(3.26) $\{\{2\}, \{3\}, \{4\}\}$

showing that $h(3,3) = 1$.

These examples contain the key for calculating a general recurrence relationship for $h(k,n)$. Focus on one *distinguished element*, denoted by e, which can either occur in a partition by itself or with other members. For instance, letting $e = 1$ in the first example above (3.24) shows that it occurs alone in partitions found in the first row and as a member with other elements in partitions found in the second row. When e appears by itself, the remaining elements must form a partition of $k-1$ sets from the remaining $n-1$ elements, which for this example corresponds to the partitions found in the second example (3.25). On the other hand, suppose that e is in a set with other members. Then there are k partitions of $n-1$ elements to which it can be added. In our example, the second row shows that 1 is added to each of the sets on the third example (3.26). Since there are k possibilities for selecting the distinguished element that number of such possibilities is given by $k\,h(k, n-1)$. These two cases count all possibilities and thus the general recurrence consists of two disjoint parts:

(3.27) $h(k,n) = \underbrace{h(k-1, n-1)}_{e \text{ is by itself}} + \underbrace{k\,h(k, n-1)}_{e \text{ is with other elements}}$

Comparing the recurrence in equation (3.18), with the recurrence just derived, equation (3.27) shows that $h(k,n) = \left\{ {n \atop k} \right\}$ and provides a combinatorial interpretation of Stirling numbers of the second kind.

A combinatoric interpretation of Stirling numbers of the first kind arises when one considers cycles in permutations. Suppose that integers 1 through n are permuted leading to $(a_1, \ldots, a_n)$. If $a_j = i$, then we say that item i in the permutation was moved to position j and represent this by $i \rightarrow j$. A *cycle* in the permutation is a sequence $i \rightarrow j \rightarrow k \rightarrow \cdots \rightarrow i$ indicating that i was moved to j, j was moved to k, and so forth until eventually the sequence returns back to i. As an example, there are three cycles for the permutation $(3, 2, 5, 6, 1, 4)$:

$$1 \rightarrow 3 \rightarrow 5 \rightarrow 1 \qquad 2 \rightarrow 2 \qquad 4 \rightarrow 6 \rightarrow 4$$

How many possible permutations are there in which k cycles are formed when n items are permuted?

To derive a recurrence for this, let $g(i, k)$ count the number of possible i-cycles when k items are permuted. Focus on one distinguished element which is added to n items. This distinguished element can either be a cycle unto itself or become part of another cycle. In the first case i cycles are created if the k other items form $i - 1$ cycles, $g(i - 1, k)$. In the second case, the distinguished item can be added to any one of the existing i cycles and this can be done by adding it to any of the k places formed by the existing items, $kg(i, k)$. These are disjoint cases and thus

$$(3.28) \qquad g(i, k + 1) = g(i - 1, k) + kg(i, k)$$

Comparing the recurrence in equation (3.6) with the recurrence just derived, equation (3.28) shows that $g(i, k+1) = \begin{bmatrix} k+1 \\ i \end{bmatrix}$ and provides a combinatorial interpretation of Stirling numbers of the first kind.

To express the recurrence relationships considered thus far in the text along with their combinatorial interpretations, let $\beta_{k,n}$ denote a recurrence relationship of n items having k sub-features (such as choices, cycles, or partitions). The following table then illustrates the differences between the primary counting regimes:

Type	Recurrence Relationship	Combinatorial Meaning
n^k	$\beta_{k,n} = n\beta_{k-1,n}$	Number of permutations of k items from a set of n with replacement
$n^{\underline{k}}$	$\beta_{k,n} = n\beta_{k-1,n-1}$	Number of permutations of k items from a set of n without replacement
$\left\langle \begin{matrix} n \\ k \end{matrix} \right\rangle$	$\beta_{k,n} = \beta_{k-1,n} \quad + \beta_{k,n-1}$	Binomial-R coefficients: The number of ways to choose k items from a set of n with replacement.
$\begin{pmatrix} n \\ k \end{pmatrix}$	$\beta_{k,n} = \beta_{k-1,n-1} + \beta_{k,n-1}$	Binomial coefficients: The number of ways to choose k items from a set of n without replacement.
$\begin{bmatrix} n \\ k \end{bmatrix}$	$\beta_{k,n} = \beta_{k-1,n-1} + (n-1)\beta_{k,n-1}$	Stirling numbers of the first kind: The number of k cycles in a permutation of n items
$\left\{ \begin{matrix} n \\ k \end{matrix} \right\}$	$\beta_{k,n} = \beta_{k-1,n-1} + k\beta_{k,n-1}$	Stirling numbers of the second kind: The number of ways to partition n items into k non-empty subsets.

Chapter 4
Fearful Symmetry

> Tyger Tyger, burning bright,
> In the forests of the night;
> What immortal hand or eye,
> Could frame thy fearful symmetry?

William Blake (1757–1827)

Symmetry might be fearful in a Tiger as William Blake alludes to
in his poem, *The Tyger*, but in mathematics it is wholly a thing of
beauty. Symmetry can often be used as a tool to cut a simple, elegant,
path through a labyrinth of mathematical obstacles. Abstractly, a
mathematical object displays the property of symmetry if it is invariant
to parametric change.

For example, a function f of n arguments is symmetric if any permutation of its arguments leaves the value of the function unchanged. To
state this precisely, let ℓ_i be a permutation of the integers 1 through n.
Then f is symmetric if it satisfies

$$f(x_1, \ldots, x_n) = f(x_{\ell_1}, \ldots, x_{\ell_n})$$

for all possible permutations. A geometric figure is said to be symmetric about an axis of symmetry if its shape is invariant to rotations
about that axis. This can be expressed in terms of the equation for
the figure. Thus if $f(x)$ is an equation for a figure on the plane, then
it is said to have even symmetry if $f(x) = f(-x)$ and odd symmetry
if $-f(x) = f(-x)$. These symmetries correspond to an invariance of
shape when the figure is rotated about a vertical line (even symmetry)
or rotated about a vertical line and a horizontal line (odd symmetry).

This chapter derives results that follow from symmetric properties
starting with simple polynomials. Classic results such as the quadratic

© Springer Nature Switzerland AG 2020   39
R. Nelson, *A Brief Journey in Discrete Mathematics*,
https://doi.org/10.1007/978-3-030-37861-5_4

formula, the equation for the line of shortest distance between a point and a line, and the Pythagorean Theorem are derived using symmetry. Expressing a polynomial in terms of its roots leads to the definition of elementary simple polynomials which in turn leads to the Newton–Giraud formula. This formula produces a wealth of identity relationships which links up symmetry with combinatorics providing a unified way to express binomial and binomial-R coefficients as well as Stirling numbers of the first and second kind.

4.1   Symmetric Functions

A function is said to be symmetric if its value is invariant under a reordering of its arguments. For example, *Euclidean* distance between two points, $p_i = (x_i, y_i)$, $i = 1, 2$ in the plane is a symmetric function of the points:

$$(4.1) \qquad d(p_1, p_2) = \sqrt{(x_2 - x_1)^2 + (y_2 - y_1)^2}$$

4.1.1   Simple Polynomials

Another example of symmetric functions arises in a first degree polynomial, a line, with equation

$$y = f(x) = ax + b$$

The slope

$$(4.2) \qquad a = \frac{y_2 - y_1}{x_2 - x_1}$$

and intercept

$$(4.3) \qquad b = \frac{y_1 x_2 - y_2 x_1}{x_2 - x_1}$$

clearly do not depend on the order of the points and thus are symmetric functions of p_1 and p_2.

A quadratic polynomial is written as

$$g(x) = ax^2 + bx + c$$

It may not be obvious from this equation but that g has a point of symmetry, which we denote by s. This implies that a parabola, the curve of a quadratic polynomial, has the same shape when flipped about the y-axis centering on point s (it has even symmetry). Consider g evaluated about the point s:

$$g(s + x) = a(s + x)^2 + b(s + x) + c$$
$$= g(s) + ax^2 + (2as + b)x$$

Notice that the coefficient of the x term above vanishes if we select s to equal

$$(4.4) \qquad\qquad\qquad s = -\frac{b}{2a}$$

The value of the function at this point is given by

$$(4.5) \qquad g(s) = \frac{b^2}{4a} - \frac{b^2}{2a} + c = c - \frac{b^2}{4a} = c - as^2$$

and the function evaluated about it by

$$(4.6) \qquad\qquad\qquad g(s + x) = g(s) + ax^2$$

Since equation (4.6) only contains an even power of x, it follows that $g(s+x) = g(s-x)$ showing that s is a point of symmetry. Since $x^2 > 0$, equation (4.6) also shows that $g(s)$ corresponds to a minimal of the function if $a > 0$ and to a maximal value if $a < 0$.

4.1.2   The Quadratic Equation

The symmetry of g affords an effortless derivation of a classical result. The *roots* of a function are places where it equals 0. In the quadratic case, to solve for the roots we can lever off our results for the symmetric point s to write

$$0 = g(s + x) = g(s) + ax^2$$

Solving this requires finding the value of $s + x$ where the equation equals 0. To determine the value of x we solve the equation yielding

$$x^2 = -\frac{g(s)}{a}$$

Taking the square root to get the x portion of the solution and then combining with s leads to a solution given by

$$x^{\star}_{1,2} = s \pm \sqrt{-\frac{g(s)}{a}}$$

Substituting (4.4) and (4.5) for s and $g(s)$, respectively, leads to the well-known *quadratic formula*

$$x^{\star}_{1,2} = \frac{-b \pm \sqrt{b^2 - 4ac}}{2a}$$

4.1.3 Equation of the Minimum Distance Line

An immediate question arises when adding another point $p_3 = (x_3, y_3)$ to the plane: what is the distance from this point to the nearest point on a line. To address this, let $y = f(x) = ax + b$ be the equation for the line and note that the distance of p_3 from the line equals 0 if p_3 lies on this line. Otherwise, to derive the minimum distance it is easier to work with square of the distance, termed the *Squared Euclidean distance*. The value of d^2 to a point (x, y) using (4.1) can be written as the quadratic

$$d^2 = (x_3 - x)^2 + (y_3 - (ax + b))^2$$

Some algebra reduces this expression to the quadratic polynomial

(4.7) $$d^2 = \alpha x^2 + \beta x + \gamma$$

where

(4.8) $\quad \alpha = 1 + m^2, \quad \beta = -2(x_3 + m(y_3 - b)), \quad \gamma = x_3^2 + (b - y_3)^2$

Let σ denote the symmetric point of this quadratic given in (4.4)

(4.9) $$\sigma = -\frac{\beta}{2\alpha}$$

Since $\alpha = 1 + a^2$ is positive, the value of the quadratic (4.7) at symmetric point α corresponds to the minimal distance. Thus the distance between point $q = (\sigma, f(\sigma))$ on the line and point $p_3 = (x_3, y_3)$ corresponds to the minimum distance. The minimal squared distance is thus given by (4.5)

$$d^2 = \gamma - \frac{\beta^2}{4\alpha} = x_3^2 + (y_3 - b)^2 - \frac{(x_3 + a(y_3 - b))^2}{1 + a^2}$$

which simplifies to

$$d^2 = \frac{(y_3 - b - ax_3)^2}{1 + a^2}$$

Taking the positive square root of this (expressed here in terms of an absolute value) shows that the minimal distance of point p_3 to the line f is given by

(4.10) $$d = \frac{|y_3 - b - ax_3|}{\sqrt{1 + a^2}} = \frac{|y_3 - f(x_3)|}{\sqrt{1 + a^2}}$$

Let the equation for the minimum distance line be given by $w(x) = ux + v$. The point of intersection on the line corresponding to the minimal distance equals $q = (\sigma, f(\sigma))$. Substituting the values for α and β in these expressions yields

(4.11) $$\sigma = \frac{x_3 + a(y_3 - b)}{1 + a^2}$$

and

(4.12) $$f(\sigma) = \frac{a^2 y_3 + ax_3 + b}{1 + a^2}$$

To determine the equations for u and v we can use the previous results derived for linear equations. The slope u of the minimal distance line, using (4.2), is given by

$$u = \frac{y_3 - f(\sigma)}{x_3 - \sigma} = \frac{y_3 - \frac{a^2 y_3 + ax_3 + b}{1 + a^2}}{x_3 - \frac{x_3 + a(y_3 - b)}{1 + a^2}}$$

which simplifies to

$$u = -\frac{1}{a}$$

The intercept, using (4.3), is given by

$$v = \frac{f(\sigma)x_3 - y_3\sigma}{x_3 - \sigma} = \frac{\frac{(a^2 y_3 + a x_3 + b)x_3 - y_3(x_3 + a(y_3 - b))}{1+a^2}}{x_3 - \frac{x_3 + a(y_3 - b)}{1+a^2}}$$

which gracefully collapses to

$$v = y_3 + \frac{x_3}{a}$$

Thus the equation for the *minimal distance line* can be compactly written as

(4.13) $$\qquad\qquad w(x) = y_3 + \frac{x_3 - x}{a}$$

4.1.4   The Pythagorean Theorem

Another well-known result depending only on symmetry arises from the previous arguments. Point $p_3 = (x_3, y_3)$ is a distance of d from point $(\sigma, f(\sigma))$ on line f. Consider another point that lies on line f denoted by $r = (x, y)$ and, to avoid a special case, assume that $r \neq q = (\sigma, f(\sigma))$. There are now three different line segments that arise when points p_3, q, and r are taken two at a time. What is the relationship between them? Once again we will deal with squared distances to simplify equations.

Equations for these three line segments can be written as follows: from equation (4.10)

$$d_{p_3,q}^2 = \frac{(y_3 - f(x_3))^2}{1 + a^2}$$

The distance between q and r can be written in terms of σ:

$$d_{q,r}^2 = (x - \sigma)^2 + (y - f(\sigma))^2$$

and between r and p by

$$d_{r,p}^2 = (x - x_3)^2 + (y - y_3)^2$$

Substituting (4.11) and (4.12) into the equation for $d_{q,r}^2$, along with tedious algebra, shows that these distances satisfy

$$d_{r,p}^2 = d_{p,q}^2 + d_{q,r}^2$$

This equation is a form of the Pythagorean Theorem.

Typical derivations of the theorem start with a triangle having one angle of 90 degrees, a *right* triangle, and then establish the equation relating the length of the hypotenuse to that of the sides of the triangle. The Pythagorean Theorem here emerged algebraically only using symmetry without requiring the notion of an angle. The obvious conclusion is that the line connecting point (x, y) along the shortest path to f intersects f in a right angle. Joining this with the Pythagorean Theorem supports the intuition that the shortest distance between two points *on a plane* is a straight line.

This conclusion, however, is a consequence of the choice of a distance metric. The Euclidean metric implies the Pythagorean Theorem whereas other metrics do not. For example, a Manhattan metric allows only movements on a grid, like the streets of Manhattan, thus allowing no diagonal moves. In contrast, a Chebyshev metric is similar to the way a King moves on a chess board which does allow diagonal moves. The shortest distance between a point and a line defined in these spaces results in different conclusions.

4.1.5 Cubic Polynomials

It makes sense at this point to ask the obvious question: does the same geometric symmetry exist for a cubic polynomial? To investigate this, write the cubic as

$$p(x) = ax^3 + bx^2 + cx + d$$

and consider the cubic around a given point s to write

$$p(s+x) = a\sum_{i=0}^{3} \binom{3}{i} s^i x^{3-i} + b\sum_{i=0}^{2} \binom{2}{i} s^i x^{2-1} + c(s+x) + d$$

$$= p(s) + (3as^2 + 2bs + c)x + (3as + b)x^2 + ax^3$$

$$= p(s) + f(x)$$

where we have defined

$$f(x) = (3as^2 + 2bs + c)x + (3as + b)x^2 + ax^3$$

Selecting a value of s so that $3as + b = 0$ eliminates the square term above

$$s = -\frac{b}{3a}$$

With minor algebra this leads to

$$f(x) = (c + bs)x + ax^3$$

and

$$p(s) = 2bs^2/3 + cs + d$$

Notice that f only has odd powers of x and thus $f(x) = -f(-x)$. This shows that p is an odd function and thus implies that the shape of the cubic is invariant to flipping it about point s while also flipping it about the x-axis:

$$f(x) = p(s+x) - p(s) = -(p(s-x) - p(s))$$

The point of symmetry for the quadratic provided a simple way to obtain its roots. Does the cubic yield its roots so easily? The roots are obtained by solving the equation

$$ax^3 + bx^2 + cx + d = 0$$

The roots do not change if the equation is divided by a. To avoid carrying around awkward notation, let the values of b, c, and d now correspond to the arguments that result after division. This creates the *monic polynomial* (a univariate polynomial where the leading coefficient equals 1)

$$x^3 + bx^2 + cx + d = 0$$

The symmetric point for this modification is given by

$$s = -\frac{b}{3}$$

Solving for the roots of this polynomial implies finding the values of x which satisfy

$$p(s + x) = f(x) + p(s) = 0$$

which, after expanding the terms, means solving

$$(c + bs)x + x^3 + p(s) = 0$$

Since $p(s)$ is a constant, the difficulty with solving this equation lies in the fact that there is a non-zero coefficient for the x term in the above equation. If it were eliminated, then taking the cube root would yield the solution.

This presents the general problem to solve

(4.14) $$x^3 + \mu x + \omega = 0$$

where μ and ω are constants. François Viète (1540–1603) proposed an ingenious substitution that reduces this problem to that of solving a quadratic. Let z be a variable that satisfies

(4.15) $$x = z + \frac{\nu}{z}$$

for some constant ν. Then substitution of (4.15) into (4.14) leads to

$$\left(z + \frac{\nu}{z}\right)^3 + \mu\left(z + \frac{\nu}{z}\right) + \omega = 0$$

Expanding and collecting terms leads to

$$z^3 + (3\nu + \mu)z + \nu(3\nu + \mu)\frac{1}{z} + \frac{\nu^3}{z^3} + \omega = 0$$

The genius of Viète's substitution is seen when the z and $1/z$ term both vanish with the selection of

$$\nu = -\frac{\mu}{3}$$

This leads to the much reduced equation

$$z^3 + \frac{\mu^3}{27z^3} + \omega = 0$$

To determine the roots for this equation, multiply by z^3 which aptly leads to a quadratic in z^3

$$\left(z^3\right)^2 + \omega \left(z^3\right) + \frac{\mu^3}{27} = 0$$

This is easily solved using the quadratic formula.

With this general solution behind us, the roots of the original polynomial can now be written by *unfolding* all of the substitutions made along the way. We will only sketch the process here. The steps include solving the quadratic, taking the cube root of the solutions to determine x, substituting the values of μ, ω, and s, including the function $p(s)$, and expressing everything in terms of b, c, and d. The path along the way poses issues that deal with imaginary numbers.

4.2   Elementary Symmetric Polynomials

The roots $r_i, i = 1, \ldots, n$, of a polynomial of degree n can be analyzed by writing the polynomial as a monic polynomial consisting of a product of terms:

$$p_n(x) = (x - r_1)(x - r_2) \cdots (x - r_n)$$
$$= x^n + a_{n-1,n}x^{n-1} + \cdots + a_{1,n}x + a_{0,n}$$

In the last equation $a_{k,n}$ denotes the coefficient of x^k ($a_{n,n} = 1$) in the expanded polynomial. Clearly $p_n(x) = 0$ whenever $x = r_i$. To obtain an idea of how to calculate the coefficients of p_n expand the first few cases:

$$p_2(x) = x^2 - (r_1 + r_2)x + r_1 r_2$$
$$p_3(x) = x^3 - (r_1 + r_2 + r_3)x^2 + (r_1 r_2 + r_1 r_3 + r_2 r_3)x - r_1 r_2 r_3$$

$$p_4(x) = x^4 - (r_1 + r_2 + r_3 + r_4)x^3 + (r_1r_2 + r_1r_3 + r_1r_4 + r_2r_3$$
$$+ r_2r_4 + r_3r_4)x^2 - -(r_1r_2r_3 + r_1r_2r_4 + r_1r_3r_4$$
$$+ r_2r_3r_4)x + r_1r_2r_3r_4$$

What is the pattern? Ignoring the sign of the coefficients for now, observe that the constant term is the product of all of the roots, the coefficient of x^{n-1} is the sum of all the roots taken one at a time, and the coefficient of x^{n-2} is the sum of all possible unique products of roots taken two at a time. In general, then, the coefficient of x^{n-k} is the sum of all possible unique product of roots taken k at a time. There are $\binom{n}{k}$ such products.

With these observations in mind, define a family of *elementary symmetric* polynomials that delineate these possibilities[1]
(4.16)
$$e_k(x_1, \ldots, x_n) = \begin{cases} \sum_{1 \le i_1 < i_2 < \cdots < i_k \le n} x_{i_1} x_{i_2} \cdots x_{i_k} & k = 1, \ldots, n, \\ 0, & k > n \end{cases}$$

In these terms, the pattern depicted above can be written as

(4.17)
$$p_n(x) = x^n + \sum_{i=1}^{n} (-1)^i e_i(r_1, \ldots, r_n) x^{n-i}$$

which compactly expresses the polynomial as a function of its roots. The coefficient of x^k of the polynomial $p_n(x)$ is thus given by

(4.18)
$$a_{k,n} = \begin{cases} 1, & k = n \\ (-1)^{n-k} e_{n-k}(r_1, \ldots, r_n), & k = 0, \ldots, n-1 \end{cases}$$

As an application of this equation consider a *falling factorial* defined by

$$n^{\underline{k}} = n(n-1) \cdots (n-k+1)$$
$$= n^k + b_{k-1,k} n^{k-1} + \ldots + b_{1,k} n$$

---

[1] This generalizes the summation found in equation (4.36) which counted the number of times an inner loop was executed in a nested set of for loops.

where the coefficients of the polynomial associated with the falling factorial are denoted by $b_{i,k}$. It is clear that there are k roots of the falling factorial polynomial that are given by $r_i = i - 1$, $i = 1, \ldots, k$. To simplify notation, let $\boldsymbol{\nu}_n$ be a vector defined by $\boldsymbol{\nu}_n = (0, 1, \ldots, n)$. Thus equation (4.18) implies that

$$(4.19) \qquad b_{i,k} = \begin{cases} 1, & i = k \\ \\ (-1)^{k-i} e_{k-i}(\boldsymbol{\nu}_{k-1}), & i = 1, \ldots, k-1 \end{cases}$$

Thus

$$(4.20) \qquad n^{\underline{k}} = \sum_{i=1}^{k} (-1)^{k-i} e_{k-i}(\boldsymbol{\nu}_{k-1}) n^i$$

To calculate an equation for the elementary symmetric polynomials in (4.20), consider the following expansion which is easily derived by mimicking the derivation of (4.17):

$$\prod_{i=1}^{n} (\lambda + z_i) = \lambda^n + \sum_{i=1}^{n} e_i(z_1, \ldots, z_n) \lambda^{n-i}$$

This equation equals 0 for $\lambda = -z_\ell$, $\ell = 1, \ldots, n$,

$$0 = (-1)^n z_\ell^n + \sum_{i=1}^{n} (-1)^{n-i} e_i(z_1, \ldots, z_n) z_\ell^{n-i}, \quad \ell = 1, \ldots, n$$

Summing all of these equations yields

$$0 = \sum_{\ell=1}^{n} \left((-1)^n z_\ell^n + \sum_{i=1}^{n} (-1)^{n-i} e_i(z_1, \ldots, z_n) z_\ell^{n-i} \right)$$

$$= (-1)^n \sum_{\ell=1}^{n} z_\ell^n + \sum_{i=1}^{n} (-1)^{n-i} e_i(z_1, \ldots, z_n) \sum_{\ell=1}^{n} z_\ell^{n-i}$$

Defining

$$(4.21) \qquad q_k(z_1, \ldots, z_n) = z_1^k + \cdots + z_n^k$$

and rewriting yields

(4.22)
$$0 = (-1)^n q_n(z_1, \ldots, z_n) + \sum_{i=1}^{n} (-1)^{n-i} e_i(z_1, \ldots, z_n) q_{n-i}(z_1, \ldots, z_n)$$

This formula, termed the *Newton–Girard formula*, was first discovered by Albert Girard (1595–1632) and later, independently, by Isaac Newton (1653–1727).

4.2.1 Newton–Girard Formula

Equation (4.22) lends itself to a recursive representation. Since the equation is valid for any number of arguments, simplify notation by dropping the argument list. Also note that $q_0 = z_1^0 + \cdots + z_n^0 = n$. Rewriting (4.22) with these simplifications, along with a reorganization of the terms, yields

(4.23)
$$n e_n = (-1)^{n+1} q_n + \sum_{i=1}^{n-1} (-1)^{n-i+1} e_i q_{n-i}$$

The first few recursions for the functions e_i are given by

$$e_1 = q_1$$
$$e_2 = \frac{1}{2}(e_1 q_1 - q_2)$$
$$e_3 = \frac{1}{3}(e_2 q_1 - e_1 q_2 + q_3)$$
$$e_4 = \frac{1}{4}(e_3 q_1 - e_2 q_2 + e_1 q_3 - q_4)$$

Unfolding these recursions leads to the equalities

(4.24)
$$e_1 = q_1$$
$$e_2 = \frac{1}{2}(q_1^2 - q_2)$$

$$e_3 = \frac{1}{6}(q_1^3 - 3q_1q_2 + 2q_3)$$

$$e_4 = \frac{1}{24}(q_1^4 - 6q_1^2q_2 + 8q_1q_3 + 3q_2^2 - 6q_4)$$

Writing a recursion for the values of q_i leads to the following equations:

$$q_1 = e_1$$
$$q_2 = e_1q_1 - 2e_2$$
$$q_3 = e_1q_2 - e_2q_1 + 3e_3$$
$$q_4 = e_1q_3 - e_2q_2 + e_3q_1 - 4e_4$$

Unfolding these recursions leads to

(4.25)
$$q_1 = e_1$$
$$q_2 = e_1^2 - 2e_2$$
$$q_3 = e_1^3 - 3e_1e_2 + 3e_3$$
$$q_4 = e_1^4 - 4e_1^2e_2 + 4e_1e_3 + 2e_2^2 - 4e_4$$

4.2.2   Identities and Combinatorial Coefficients

With these recursions in place it is easy to compute the values e_k for the falling factorial considered in equation (4.20) (closed form expressions for these values can also be written). Note that

(4.26) $$q_k(\boldsymbol{\nu}_{\ell-1}) = S_{k,\ell-1}$$

where S is the sum of integer powers defined by

$$S_{k,n} = 1 + 2^k + \cdots + n^k$$

Substituting this into (4.23) for the first few values yields the following table for the values of e_k:

	Table of $e_k(\boldsymbol{\nu}_n)$						
			n				
$k\backslash n$	1	2	3	4	5	6	7
1	1	0	0	0	0	0	0
2	3	2	0	0	0	0	0
3	6	11	6	0	0	0	0
4	10	35	50	24	0	0	0
5	15	85	225	274	120	0	0
6	21	175	735	1,624	1,764	720	0
7	28	322	1,960	6,769	13,132	13,068	5,040

As an example of using the results of this table consider the third row in the table which shows

$$n^{\underline{4}} = n^4 - 6n^3 + 11n^2 - 6n$$

This table is similar to that of page 31 associated with Stirling numbers of the first kind (page 29). Linking equation (4.20) with the derivation of equation (3.13) in that chapter establishes the relationship between these two mathematical systems

$$(4.27) \qquad n^{\underline{k}} = \sum_{i=1}^{k}(-1)^{k-i}e_{k-i}(\boldsymbol{\nu}_{k-1})n^i = \sum_{i=1}^{k}(-1)^{k-i}\begin{bmatrix}k\\i\end{bmatrix}n^i$$

Matching exponents of n establishes an equation between elementary symmetric polynomials and Stirling numbers of the first kind:

$$(4.28) \qquad e_{k-i}(\boldsymbol{\nu}_{k-1}) = \begin{bmatrix}k\\i\end{bmatrix}, \quad i = 1,\ldots,k-1$$

The Newton–Girard formula opens up a literal ocean of possible identities . To state a few, use (4.26) and (4.28) in the equations for e_i in (4.24) along with some minor index manipulations, to create

$$(4.29) \qquad\qquad\qquad\qquad S_{1,\ell} = \begin{bmatrix}\ell+1\\\ell\end{bmatrix}$$

$$\frac{1}{2}\left(S_{1,\ell}^2 - S_{2,\ell}\right) = \begin{bmatrix}\ell+1\\\ell-1\end{bmatrix}$$

$$\frac{1}{6}\left(S_{1,\ell}^3 - 3S_{1,\ell}S_{2,\ell} + 2S_{3,\ell}\right) = \begin{bmatrix} \ell+1 \\ \ell-2 \end{bmatrix}$$

$$\frac{1}{24}\left(S_{1,\ell}^4 - 6S_{1,\ell}^2 S_{2,\ell} + 8S_{1,\ell}S_{3,\ell} + 3S_{2,\ell}^2 - 6S_{4,\ell}\right) = \begin{bmatrix} \ell+1 \\ \ell-3 \end{bmatrix}$$

To proceed along the complementary path, use the equations for q_i in (4.25) to produce the identities

(4.30)
$$S_{2,\ell} = \begin{bmatrix} \ell+1 \\ \ell \end{bmatrix}^2 - 2\begin{bmatrix} \ell+1 \\ \ell-1 \end{bmatrix}$$

(4.31)
$$S_{3,\ell} = \begin{bmatrix} \ell+1 \\ \ell \end{bmatrix}^3 - 3\begin{bmatrix} \ell+1 \\ \ell \end{bmatrix}\begin{bmatrix} \ell+1 \\ \ell-1 \end{bmatrix} + 3\begin{bmatrix} \ell+1 \\ \ell-2 \end{bmatrix}$$

To consider the special case where $z_i = 1$, $i = 1, \ldots, n$, let $\mathbf{1}_n$ be a length n vector of 1's, $\mathbf{1}_n = \underbrace{(1, \ldots, 1)}_{n \text{ times}}$ and note that

(4.32)
$$q_k(\mathbf{1}_n) = n, \quad k = 1, \ldots$$

Substituting into (4.24) yields

(4.33)
$$e_1(\mathbf{1}_n) = n$$

$$e_2(\mathbf{1}_n) = \frac{1}{2}(n^2 - n)$$

$$e_3(\mathbf{1}_n) = \frac{1}{6}(n^3 - 3n^2 + 2n)$$

$$e_4(\mathbf{1}_n) = \frac{1}{24}(n^4 - 6n^3 + 11n^2 - 6n)$$

Equation (4.27) implies that we can write this pattern by

(4.34)
$$e_k(\mathbf{1}_n) = \frac{n^{\underline{k}}}{k!} = \binom{n}{k}$$

Substituting this into (4.25) generates the following sequence of identities:

$$(4.35) \qquad n = n^2 - 2 \binom{n}{2}$$

$$= n^3 - 3n \binom{n}{2} + 3 \binom{n}{3}$$

$$= n^4 - 4n^2 \binom{n}{2} + 4n \binom{n}{3} + 2 \binom{n}{2}^2 - 4 \binom{n}{4}$$

Unfolding (4.34) yields an identity pertinent to computer programming that counts the number of times the inner loop is executed in a nested series of for loops

$$(4.36) \qquad \sum_{1 \le i_1 < \cdots < i_k \le n} 1 = \binom{n}{k}$$

4.2.3 Inclusion–Exclusion

A novel use of identity (4.36) is to determine the size of a set. To explain this, let $|x|$ denote the number of elements of a set x and let $a_i, i = 1, \ldots n$ be a family of sets.. For $n = 2$ a familiar result shows that

$$(4.37) \qquad |a_1 \cup a_2| = |a_1| + |a_2| - |a_1 \cap a_2|$$

The analogous statement for $n = 3$ is

$$(4.38)$$
$$|a_1 \cup a_2 \cup a_3|$$
$$= |a_1| + |a_2| + |a_3| - |a_1 \cap a_2| - |a_1 \cap a_3| - |a_2 \cap a_3| + |a_1 \cap a_2 \cap a_3|$$

The *inclusion–exclusion* principle generalizes these results and states that

(4.39)

$$|a_1 \cup \cdots \cup a_n|$$

$$= \sum_{i=1}^{n} |a_i| - \sum_{1 \leq i_1 < i_2 \leq n} |a_{i_1} \cap a_{i_2}| + \sum_{1 \leq i_1 < i_2 < i_3 \leq n} |a_{i_1} \cap a_{i_2} \cap a_{i_3}| +$$

$$\cdots + (-1)^{n-1} |a_1 \cap \cdots \cap a_n|$$

To prove this, note that if element α is in any one of the sets a_i, then the left-hand side of equation (4.39) equals 1. Suppose then that α is contained in ℓ sets and, without loss of generality, assume that these are the sets 1 through ℓ. Thus $\alpha \in a_i, i = 1, \ldots, \ell$ and $\alpha \notin a_j, j = \ell + 1, \ldots, n$. This implies that intersections on the right-hand side of (4.39) with any set a_i where $i > \ell$ add 0 to the equation. Concentrating on a particular intersection on the right-hand side shows that

$$\sum_{1 \leq i_1 < \cdots < i_k \leq n} |a_{i_1} \cap \cdots \cap a_{i_k}| = \sum_{1 \leq i_1 < \cdots < i_k \leq \ell} |a_{i_1} \cap \cdots \cap a_{i_k}|$$

$$= \sum_{1 \leq i_1 < \cdots < i_k \leq \ell} 1$$

$$= \binom{\ell}{k}$$

where we have used the loop counting identity (4.36) in the last simplification. Thus the right-hand side of equation (4.39) for element α yields

$$\ell - \binom{\ell}{2} + \binom{\ell}{3} + \cdots + (-1)^{\ell-1} \binom{\ell}{\ell} = \sum_{i=1}^{\ell} (-1)^{i-1} \binom{\ell}{i} = 1$$

where the last equality comes from identity (2.28).

Often, when equations that have alternating signs, it is an indication that they can be derived using inclusion–exclusion. One case where this previously occurred involved dearrangements, equation (2.10). The form of this equation correctly suggests that an inclusion–exclusion argument could have been used to derive the result.

4.3 Fundamental Theorem of Symmetric Polynomials

There is a deeper theorem underlying the previous results. Both the elementary symmetric polynomials (4.16) and the power sum polynomials (4.21) are specific examples of symmetric polynomial functions. Other examples include

$$F(X_1, X_2) = X_1^2 X_2^2 - X_1^3 X_2 - X_1 X_2^3$$
$$F(X_1, X_2, X_3) = X_1 X_2 X_3 + X_1 X_2 + X_2 X_3 + X_1 X_3$$
$$F(X_1, X_2, X_3) = (X_1 - X_2)^2 + (X_1 - X_3)^2 + (X_2 - X_3)^2$$

There are, however, many other families of polynomials that are symmetric. For example, the *complete homogeneous symmetric polynomials* are defined similarly to the elementary symmetric polynomials by replacing the $<$ sign in the summation by a $\leq$ sign: (4.40)

$$h_k(x_1, \ldots, x_n) = \begin{cases} \sum_{1 \leq i_1 \leq i_2 \leq \cdots \leq i_k \leq n} x_{i_1} x_{i_2} \cdots x_{i_k} & k = 1, \ldots, n, \\ 0, & k > n \end{cases}$$

Similar to equation (4.34), it can be shown that binomial-R coefficients satisfy

$$(4.41) \qquad h_k(\mathbf{1}_n) = \left\langle \begin{matrix} n \\ k \end{matrix} \right\rangle$$

In computer programming, this counts the number of loops where in (4.36) less than signs are replaced with less than or equal signs can be derived which shows that

$$(4.42) \qquad \sum_{1 \leq i_1 \leq \cdots \leq i_k \leq n} 1 = \left\langle \begin{matrix} n \\ k \end{matrix} \right\rangle$$

It is also the case that, like equation (4.28) for Stirling numbers of the first kind, Stirling numbers of the second kind can be expressed in terms of the family of homogeneous symmetric polynomials:

$$(4.43) \qquad h_{n-k}(\boldsymbol{\nu}_k) = \left\{ \begin{matrix} n \\ k \end{matrix} \right\}$$

The results linking combinatorial coefficients with symmetric polynomials are summarized in the table below:

Type	Combinatorial Term	Symmetry
$\binom{n}{k}$	Binomial Coefficient	$e_k(\mathbf{1}_n)$
$\left\langle \begin{matrix} n \\ k \end{matrix} \right\rangle$	Binomial-R Coefficient	$h_k(\mathbf{1}_n)$
$\left[\begin{matrix} n \\ k \end{matrix} \right]$	Stirling Numbers of the First Kind	$e_{n-k}(\boldsymbol{\nu}_{n-1})$
$\left\{ \begin{matrix} n \\ k \end{matrix} \right\}$	Stirling Numbers of the Second Kind	$h_{n-k}(\boldsymbol{\nu}_k)$

The *monomial symmetric polynomials* are another family of symmetric polynomials which are specified through exponents that correspond to a partition of an integer. An example can be used as a quick way to define these polynomials. Consider a partition of the integer 6 given by $(3, 2, 1)$, $6 = 3 + 2 + 1$ then

$$m_{(3,2,1)} = X_1^3 X_2^2 X_3 + X_1^3 X_2 X_3^2 + X_1^2 X_2^3 X_3 + X_1^2 X_2 X_3^3$$
$$+ X_1 X_2^3 X_3^2 + X_1 X_2^2 X_3^3$$

The partition of the integer k given by

$$\boldsymbol{\lambda}_k = (\underbrace{1, \ldots, 1}_{k \text{ 1's}}, \underbrace{0, \ldots, 0}_{n-k \text{ 0's}})$$

shows that the elementary symmetric polynomials and the monomial symmetric polynomials are related by

$$(4.44) \qquad e_k(x_1, \ldots, x_n) = m_{\boldsymbol{\lambda}_k}(x_1, \ldots, x_n)$$

There are many other examples of families of symmetric polynomials but the four examples defined so far are sufficient to make our point. To aid in this, it suffices to simply state two equations that are similar in spirit to the Newton–Girard formula (4.22):

$$(4.45) \qquad 0 = \sum_{k=0}^{n} (-1)^i e_k(x_1, \ldots, x_n) h_{n-k}(x_1, \ldots, x_n)$$

and

$$(4.46) \quad 0 = \sum_{\lambda=\text{all partitions of integer } k} m_\lambda(x_1, \ldots, x_n) - h_k(x_1, \ldots, x_n)$$

Something deeper is at hand. Equations (4.22), (4.44), (4.45), and (4.46) all depict a weighted summation that equals 0 that relates to the elementary symmetric polynomials. These equations all point towards the same source: *the Fundamental Theorem of Symmetric Polynomials*. This theorem states that any symmetric polynomial can be expressed as a unique combination of the elementary symmetric polynomials. The families of symmetric polynomials defined so far, the power sum, the homogeneous, and the monomial, are all special cases of this theorem.

This chapter has now almost come close to full closure. It started with the symmetry found in an Euclidean distance metric and with the equations for the slope and intersect of a linear equation. This led to the equation for the minimal distance from a point to the line which required the symmetry of a quadratic polynomial. A byproduct of this symmetry was the quadratic formula and the Pythagorean Theorem. As an aside cubic polynomials were shown to have a reverse symmetry. The elementary symmetric polynomials were derived to analyze the structure of the roots of a polynomial and this led to an equation between these polynomials and the power sum polynomials. A wealth of identities were derived leading the way to the fundamental theorem of symmetric polynomials.

4.4 Galois' Theorem and Numerical Solutions

One of mathematics more bizarre stories deals with the impossibility of a general closed form equation for the roots of polynomials of degree 5 or higher. This was proved by Évariste Galois (1811–1832) and, at first, was not accepted by the mathematical community. Under premonitions of his impending death, Galois wrote a mathematical testament the night before a scheduled duel. In the margins of the manuscript he wrote the foreboding words, *"There is something to complete in this demonstration. I do not have the time"*. It is presumed that the duel had something to do with his love relationship, or lack thereof, with Stephanie-Felice du Motel. Although his results eventually won the minds of mathematicians and led to a branch of mathematics called *Galois Theory*, he lost in the heart of love and was wounded in the duel. Pitilessly abandoned on the field of battle by his opponent, a peasant found him and he died the next day at the young age of 20.

If you cannot write an equation for the solution of the roots of a higher order polynomial, then a numerical technique must be utilized. One way to create such an algorithm is through a series of linear approximations. To describe one such algorithm, let $f(x)$ be a function and let x_0 and x_1 be two different values that are selected to be close to a root.[2] If either $f(x_0)$ or $f(x_1)$ equals 0 then a root has been found and the algorithm stops. The line connecting these points has a slope given by equation (4.2)

$$m_1 = \frac{f(x_1) - f(x_0)}{x_1 - x_0}$$

and an intercept given by equation (4.3)

$$h_1 = \frac{x_1 f(x_0) - x_0 f(x_1)}{x_1 - x_0}$$

The root of this linear equation, given by the next iterate x_2, is an estimate for a root of f. This is found by solving

$$0 = m_1 x_2 + h_1$$

which implies that

(4.47) $$x_2 = \frac{-h_1}{m_1} = \frac{x_0 f(x_1) - x_1 f(x_0)}{f(x_1) - f(x_0)}$$

If function f is linear over the range of x_0 and x_1, then x_2 is a root and the algorithm stops, otherwise the iteration continues by creating a series of estimates given by a straightforward index modification of (4.47):

$$x_i = \frac{x_{i-2} f(x_{i-1}) - x_{i-1} f(x_{i-2})}{f(x_{i-1}) - f(x_{i-2})}, \qquad i = 2, \ldots$$

---

[2]Selecting initial points to start the algorithm is often an art rather than a science. It is not uncommon to perform a set of experiments with a variety of initial points.

This algorithm, termed the *sequent method*, corresponds to a series of linear approximations that continue until $|x_{n+1} - x_n|$ becomes sufficiently small or a limit on the number of iterations is exceeded.

The basic idea of the algorithm is that as the x_{i-1} and x_i get closer, the function for that small range is closely approximated by a straight line. Typically this assumption holds and the algorithm converges to a root. It is possible, however, to start with a poor choice of initial points, or to have a function that is difficult to approximate linearly even over small ranges. In such cases, the algorithm stops after the iteration limit is achieved without finding a root.

To illustrate the algorithm, consider the quintic rising factorial polynomial

$$x^{\overline{5}} = x(x + 1)(x + 2)(x + 3)(x + 4) = x^5 + 10x^4 + 35x^3 + 50x^2 + 24$$

which obviously has roots $-i$ for $i = 0, 1, 2, 3, 4$. Starting with $x_0 = -2.1$ and $x_1 = -3.9$ yields the following output showing that convergence to the root at $x = -3$ occurs after six iterations

i	x_i	$f(x_i)$
0	-2.1	-0.39501
1	-3.9	1.93401
2	-2.40528634361233	-1.29922451355076
3	-3.00591355298394	0.0356551283383601
4	-2.9898705783504	-0.0602583598595037
5	-2.99994969607251	-0.000301810911873649
6	-3.00000043261029	2.59566265598619e-06
7	-2.99999999998186	-1.08816955444756e-10
8	-3	0

The field of *numerical analysis* launches from this platform.

Chapter 5
All That Glitters Is Not Gold

All that glisters is not gold;
Often have you heard that told:
Many a man his life hath sold
But my outside to behold:
Gilded tombs do worms enfold.

William Shakespeare (1564–1616)
The Merchant of Venice, Act 2, scene 7

Despite the temptations of gold alluded to in Shakespeare's verse above from *The Merchant of Venice*, the pursuit of *mathematical gold* leads, not to gilded tombs, but to the paradise of the Elysian fields of ancient Greece. Our journey in this chapter takes us back to the days of Phidias (480–430 BC), a Greek sculptor and mathematician who is said to have helped with the design of the Parthenon. The approach in this chapter uses a simple artifice—the ratio of two line segments.

5.1   The Golden Ratio

Consider a line consisting of two line segments. The first segment has a length of 1 unit and the other has length $1 - \epsilon$ where $0 < \epsilon < 1$. By construction, the second length is the smaller of the two. We are going to select the value of ϵ that equalizes two ratios. The first ratio, r_1, is the length of the total line segment to that of the larger section, thus

$$(5.1) \qquad r_1 = \frac{1 + 1 - \epsilon}{1} = 2 - \epsilon$$

© Springer Nature Switzerland AG 2020
R. Nelson, *A Brief Journey in Discrete Mathematics*,
https://doi.org/10.1007/978-3-030-37861-5_5

The second ratio, r_2, is the length of the larger segment to that of the smaller segment

$$(5.2) \qquad\qquad r_2 = \frac{1}{1 - \epsilon}$$

For some value $\epsilon^\star$ the two ratios are equal. It is not hard to derive an equation for $\epsilon^\star$:

$$2 - \epsilon^\star = \frac{1}{1 - \epsilon^\star}$$

or that

$$(\epsilon^\star)^2 - 3\epsilon^\star + 1 = 0$$

Using the quadratic formula yields the solution to this quadratic polynomial

$$\epsilon^\star = \frac{3 \pm \sqrt{5}}{2}$$

The solution corresponding to the positive square root is larger than 1 which lies outside the bound on the length of the second segment so[1]

$$\epsilon^\star = \frac{3 - \sqrt{5}}{2} = 0.381966011250105\ldots$$

The value of $\epsilon^\star$ can be calculated using (5.1) and shows that

$$(5.3)\ r_1 = 2 - \epsilon^\star = 2 - \frac{3 - \sqrt{5}}{2} = \frac{1 + \sqrt{5}}{2} = \phi = 1.61803398874989\ldots$$

This calculated value, typically denoted by ϕ, is the *golden ratio* that was admired by Phidias who used it in the design of the shape of the Parthenon.[2]

Equation (5.2) expresses ϕ differently

---

[1] The appearance of $\sqrt{5}$ implies that $\epsilon^\star$ is irrational, see the proof on page 171.
[2] The Internet is replete with interesting historical facts dealing with this ratio.

$$(5.4) \qquad r_2 = \frac{1}{1 - \epsilon^\star} = \frac{1}{1 - \frac{3-\sqrt{5}}{2}} = \frac{1}{\frac{-1+\sqrt{5}}{2}} = \frac{2}{\sqrt{5} - 1} = \phi$$

These two ratios provide two equations for the golden ratio. Straight-forward algebra shows that

$$(5.5) \qquad \phi = \frac{1 + \sqrt{5}}{2} \qquad \Longrightarrow \qquad \sqrt{5} = 2\phi - 1$$

Using this in (5.4) yields

$$\phi = \frac{2}{\sqrt{5} - 1} = \frac{2}{2\phi - 2} = \frac{1}{\phi - 1}$$

and thus

$$\phi^2 = \phi + 1$$

This shows that the golden ratio is *one of the solutions* to the quadratic equation $x^2 = x + 1$, an equation which, in this chapter, is termed *the defining equation*. A key observation to make about this equation is that the left-hand side, x^2, corresponds to a multiplication whereas the right-hand side, $x + 1$, is an addition. In essence, the equation converts multiplication to addition. What can such an observation yield?

5.1.1 Fibonacci Numbers

Since ϕ is one solution to the defining equation this means all occurrences of ϕ^2 can be replaced with $\phi + 1$ without changing the value of an expression. This can be used to calculate an expression for ϕ^3:

$$\phi^3 = \phi\phi^2 = \phi(\phi + 1) = \phi^2 + \phi = \phi + 1 + \phi = 2\phi + 1$$

Using this to calculate the next power shows that

$$\phi^4 = \phi\phi^3 = \phi(2\phi + 1) = 2\phi^2 + \phi = 2(\phi + 1) + \phi = 3\phi + 2$$

Continuing with this progression yields

$$\phi^5 = \phi\phi^4 = \phi(3\phi + 2) = 3\phi^2 + 2\phi = 3(\phi + 1) + 2\phi = 5\phi + 3$$

$$\phi^6 = \phi\phi^5 = \phi(5\phi + 3) = 5\phi^2 + 3\phi = 5(\phi + 1) + 3\phi = 8\phi + 5$$

$$\phi^7 = \phi\phi^6 = \phi(8\phi + 5) = 8\phi^2 + 5\phi = 8(\phi + 1) + 5\phi = 13\phi + 8$$

$$\phi^8 = \phi\phi^7 = \phi(13\phi + 8) = 13\phi^2 + 8\phi = 13(\phi + 1) + 8\phi = 21\phi + 13$$

The numbers have an intriguing progression and to see the pattern more clearly consider the following table:

Power	Multiple of ϕ	Constant
1	1	0
2	1	1
3	2	1
4	3	2
5	5	3
6	8	5
7	13	8
8	21	13

The first thing to note is that the multiplier equals the constant for the following power. In other words the values are just shifted versions of each other. There is another observation that comes from looking at the sequence of numbers

$$1, \; 1, \; 2, \; 3, \; 5, \; 8, \; 13, \; 21, \ldots$$

Observe that after starting with two, 1's, the next number is the sum of the previous 2. This sequence of numbers is the famous *Fibonacci sequence* named for the Italian mathematician, Leonardo of Pisa (1170–1240?) around the year 1200. The Internet is replete with the history and myriad applications of this sequence including the mating characteristics of rabbits.

Let f_i, $i = 0, \ldots$ denote the Fibonacci numbers so that $f_1 = 1$, $f_2 = 1$, $f_3 = 2$, $f_4 = 5$, and $f_n = f_{n-1} + f_{n-2}$, $n = 4, \ldots$. It is customary to start the sequence with $f_0 = 0$. The above table suggests that

$$(5.6) \qquad \phi^n = f_n \phi + f_{n-1}, \quad n = 2, \ldots$$

To prove this, use the technique of substituting $\phi + 1$ for all occurrences of ϕ^2:

$$\phi^{n+1} = \phi\phi^n = \phi(f_n\phi + f_{n-1})$$

$$= f_n\phi^2 + \phi f_{n-1} = f_n(\phi + 1) + \phi f_{n-1} = (f_n + f_{n-1})\phi + f_n$$

Since $f_{n+1} = f_n + f_{n-1}$ the last expression can be rewritten as

$$(5.7) \qquad \phi^{n+1} = f_{n+1}\phi + f_n$$

which shows that the pattern continues to the $n + 1$'st case. There is an analogy in the recurrence relationship of the Fibonacci numbers with the golden ratio seen by writing patterns side by side:

$$f_n = f_{n-1} + f_{n-2} \qquad \phi^n = \phi^{n-1} + \phi^{n-2}$$

To derive the second expression, write

$$\phi^n = \phi^{n-2}\phi^2 = \phi^{n-2}(\phi + 1) = \phi^{n-1} + \phi^{n-2}, \quad n = 3, \ldots$$

5.1.2 A Closed Form Solution

To derive a closed form equation for f_n, return to the other solution besides ϕ to the defining quadratic equation $x^2 = x + 1$. The second solution from the quadratic formula is

$$(5.8) \qquad \psi = \frac{1 - \sqrt{5}}{2} = -0.61803398874989 \ldots$$

This solution necessarily shares properties similar to ϕ since it also satisfies $\psi^2 = \psi + 1$. This implies that the pattern derived above for powers of ϕ also holds for powers of ψ

(5.9) $\psi^n = f_n \psi + f_{n-1}, \quad n = 2, \ldots$

This equation gives the key in finding an expression for f_n. Subtract ψ^n of equation (5.9) from ϕ^n of (5.6) to get:

$$\phi^n - \psi^n = f_n \phi + f_{n-1} - (f_n \psi + f_{n-1}) = f_n (\phi - \psi)$$

and, like picking a rabbit out of hat, this shows that

$$f_n = \frac{\phi^n - \psi^n}{\phi - \psi}$$

This almost seems too easy a way to get such a difficult result. This expression can be simplified since

$$\phi - \psi = \frac{1 + \sqrt{5}}{2} - \frac{1 - \sqrt{5}}{2} = \sqrt{5}$$

which yields the compact formula

(5.10) $f_n = \dfrac{\phi^n - \psi^n}{\sqrt{5}}$

This equation was first derived along different lines by Jacques Binet (1786–1856).

    Before ending this section, observe that substitution (5.5) into equation (5.8) shows that

$$\psi = \frac{1 - \sqrt{5}}{2} = \frac{1 - (2\phi - 1)}{2} = \frac{2 - 2\phi}{2} = 1 - \phi$$

which yields

$$\phi + \psi = 1$$

This is useful in deriving an expression for the sum of ϕ^n and ψ^n. Using equations (5.7) and (5.9) yields

$$\phi^n + \psi^n = f_n \phi + f_{n-1} + (f_n \psi + f_{n-1})$$
$$= f_n (\phi + \psi) + 2f_{n-1} = f_n + 2f_{n-1}$$

This can be rewritten in an easier form

(5.11) $\phi^n + \psi^n = f_n + 2f_{n-1} = f_n + f_{n-1} + f_{n-1} = f_{n+1} + f_{n-1}$

which leads to the sequence of numbers[3]:

(5.12) $2, \ 1, \ 3, \ 4, \ 7, \ 11, \ 18, \ 29, \ldots$

The Fibonacci recurrence relationship invites one to derive identities between the values. For example, write the equation for f_{2n}

$$f_{2n} = \frac{\phi^{2n} - \psi^{2n}}{\sqrt{5}} = \frac{(\phi^n)^2 - (\psi^n)^2}{\sqrt{5}}$$

and use the identity $x^2 - y^2 = (x - y)(x + y)$ and the equation for $\phi^n + \psi^n$ above in (5.11) to obtain

(5.13) $f_{2n} = \dfrac{\phi^n - \psi^n}{\sqrt{5}} (\phi^n + \psi^n) = f_n(f_{n+1} + f_{n-1})$

To derive the same value along a different line write

$$f_{2n} = \frac{\phi^{2n} - \psi^{2n}}{\sqrt{5}} = \frac{(\phi^2)^n - (\psi^2)^n}{\sqrt{5}}$$

Use the fact that $\phi^2 = 1 + \phi$ and $\psi^2 = 1 + \psi$, and the binomial theorem to express f_{2n} as
(5.14)

$$f_{2n} = \frac{(1 + \phi)^n - (1 + \psi)^n}{\sqrt{5}} = \sum_{k=0}^{n} \binom{n}{k} \frac{\phi^k - \psi^k}{\sqrt{5}} = \sum_{k=0}^{n} \binom{n}{k} f_k$$

Equating the two identities yields the relationships

$$f_{2n} = f_n(f_{n+1} + f_{n-1}) = \sum_{k=0}^{n} \binom{n}{k} f_k$$

---

[3]This sequence is termed the Lucas sequence and will be revisited later in the chapter, see equation (5.23).

Equation (5.14) corresponds to a binomial expansion of the values f_k. Using binomial inversion thus leads to another identity involving Fibonacci numbers:

$$(5.15) \qquad f_n = \sum_{k=0}^{n} \binom{n}{k} (-1)^{n-k} f_{2k}$$

To consider another identity, let $\alpha_n = f_{n+1} f_{n-1} - f_n^2$ and observe that $\alpha_1 = -1$ and $\alpha_2 = 1$. This suggests that $\alpha_n = (-1)^2$. The following set of manipulations shows that

$$\begin{aligned} \alpha_n &= f_{n+1} f_{n-1} - f_n^2 \\ &= (f_n + f_{n-1}) f_{n-1} - f_n^2 \\ &= f_{n-1}^2 - f_n(f_{n-1} + f_{n-2} - f_{n-1}) \\ &= f_{n-1}^2 - f_n f_{n-1} \\ &= -\alpha_{n-1} \end{aligned}$$

which thus yields the identity

$$(5.16) \qquad f_{n+1} f_{n-1} - f_n^2 = (-1)^n$$

One more identity can easily be obtained. Consider the product of two successive Fibonacci numbers

$$f_i f_{i-1} = (f_{i-1} + f_{i-2}) f_{i-1} = f_{i-1}^2 + f_{i-1} f_{i-2}$$

This recursive equation is easily solved, leading to the identity

$$(5.17) \qquad f_i f_{i-1} = \sum_{j=1}^{i-1} f_j^2, \quad i = 1, \ldots$$

The process of creating identities for Fibonacci numbers could go on almost endlessly since there are literally tens of thousands of such relationships.

The equation $\phi^2 = \phi + 1$ leads to some beautiful equations. Take the square root of this to yield

$$\phi = \sqrt{\phi^2} = \sqrt{\phi + 1}$$

Substituting the square root expression for every occurrence of ϕ shows that:

$$\phi = \sqrt{1 + \phi} = \sqrt{1 + \sqrt{1 + \phi}} = \sqrt{1 + \sqrt{1 + \sqrt{1 + \phi}}}$$

Clearly this continues without end and thus ϕ arises from the infinite cascade of square roots:

(5.18)
$$\phi = \sqrt{1 + \sqrt{1 + \sqrt{1 + \sqrt{1 + \cdots}}}}$$

If the defining equation is rewritten as

$$\phi = 1 + \frac{1}{\phi}$$

then continual substitution for ϕ leads to an infinite continued fraction expansion of ϕ^4:

(5.19)
$$\phi = 1 + \cfrac{1}{1 + \cfrac{1}{1 + \cfrac{1}{1 + \cfrac{1}{1 + \cfrac{1}{\ddots}}}}}$$

The golden ratio $\phi \approx 1.6180339$ is one of a handful of fundamental constants like $\pi \approx 3.1415926$, the base of the natural logarithm $e \approx 2.7182818$, or Euler's constant $\gamma \approx 0.5772156$.

---

[4] An alternative derivation is found in (10.14).

5.2   An Alternate Derivation

Viewing Fibonacci numbers through the eyes of a different model reveals a strikingly different closed form expression. Consider the number of possible sequences of the integers 1 or 2 that, when summed, equals n. Let $\rho(n)$ be the number of different ways to do this. For example, $\rho(2) = 2$ since the sequences 11 and 2 are the only possibilities. Other cases include $\rho(3) = 3$, (111, 12 and 21) and $\rho(4) = 5$, (1111, 112, 121, 211, and 22).

Comparing these values to Fibonacci numbers suggests that $\rho(n) = f_{n+1}$. To show that this is the case it suffices to establish the recurrence

$$(5.20) \qquad\qquad \rho(n) = \rho(n-1) + \rho(n-2)$$

The small examples above show that this is satisfied for values of $n \leq 4$. The equation is established by induction. For $n > 4$, consider a sequence that adds up to n that starts with a 1. In this case, the remaining integers of the sequence necessarily add up to $n-1$ which we know amounts to $\rho(n-1)$ possible sequences. Similarly for those sequences that start with 2 the remaining numbers must add to $n-2$ which equals a total of $\rho(n-2)$ sequences. These two cases account for all possibilities and thus establishes (5.20).

This perspective opens up a different form of a closed form expression for $\rho(n)$ and thus also for f_{n+1}. Partition all sequences of 1s and 2s by their length. There is only one sequence of length n which consists of all 1s. Consider a sequence having i, 2s, in it. Such a sequence has $n - 2i \geq 0$, 1s, since the sum equals n. This implies the sequence length is $n - 2i + i = n - i$. There are $\binom{n-i}{i}$ ways of choosing the places for the i, 2s, in such sequences. Since $n - 2i > 0$, it must be the case that $i \leq \lfloor n/2 \rfloor$ where $\lfloor x \rfloor$ is the integer portion of x. Summing all possibilities shows that

$$(5.21) \qquad \rho(n) = \sum_{i=0}^{\lfloor n/2 \rfloor} \binom{n-i}{i} \qquad\qquad \left(f_{n+1} = \frac{\phi^{n+1} - \psi^{n+1}}{\sqrt{5}} \right)$$

In equation (5.21) the previous closed form expression (5.10) is shown to contrast the striking difference between the two derivations. Such contrasts often happen in mathematics in the form of equations that arise for the same mathematical quantity when viewed from different perspectives. Adjusting the indices of (5.21) to be more natural provides the following identity:

$$(5.22) \qquad f_{2n+1} = \sum_{i=0}^{n} \binom{n+i}{2i} = \sum_{i=0}^{n} \binom{n+i}{n-i}$$

5.3 Generalized Fibonacci Numbers

Clearly the starting values for the Fibonacci series are arbitrary and have little influence on any basic properties derived from the essential recurrence $f_n = f_{n-1} + f_{n-2}$. For example, if the sequence starts with a, b (they both cannot be 0) then these *generalized Fibonacci numbers* satisfy

$$g_n = a f_{n-1} + b f_n$$

The widely studied *Lucas* sequence (named after Édouard Lucas (1842–1891)), that arises with the selection $a = 2$ and $b = 1$, yields[5]

$$(5.23) \qquad 2, 1, 3, 4, 7, 11, 18, 29, 47, 76, 123, 199, \ldots$$

A closed form solution for g_n is a modification of equation (5.10)

$$
\begin{aligned}
g_n &= a \left(\frac{\phi^{n-1} - \psi^{n-1}}{\sqrt{5}} \right) + b \left(\frac{\phi^n - \psi^n}{\sqrt{5}} \right) \\
&= \frac{\phi^n (b + a\phi^{-1}) - \psi^n (b + a\psi^{-1})}{\sqrt{5}} \\
&= \frac{\phi^n (b - a\psi) - \psi^n (b - a\phi)}{\sqrt{5}}
\end{aligned}
$$

The last simplification uses the equation $\phi\psi = -1$.

---

[5] Equation (5.11) also generated this sequence.

5.4   k-Bonacci Numbers

Recall that Fibonacci numbers arise from the defining equation $x^2 = x + 1$. Consider the next highest defining equation given by $x^3 = x^2 + x + 1$. Mimicking the previous derivation of Fibonacci numbers in Section 5.1.1 by deriving powers of x yields the following pattern:

Power	x^2 term	x term	1 term
x^3	1	1	1
x^4	2	2	1
x^5	4	3	2
x^6	7	6	4
x^7	13	11	7

Notice that the numbers in each row are the sum of the previous *three* numbers in the preceding rows. Such numbers are said to be *Tribonacci numbers* and the differences in sequences shown in the columns arise, like the difference between Fibonacci and Lucas numbers, from their different initial values. In particular, the first and third column correspond to the sequence starting with $0, 0, 1$,

$$0, 0, 1, 1, 2, 4, 7, 13, \cdots$$

whereas the second column corresponds to the sequence starting with $0, 1, 0$,

$$0, 1, 0, 1, 2, 3, 6, 11, \cdots$$

In all cases, the general term of the recurrence satisfies $t_n = t_{n-1} + t_{n-2} + t_{n-3}$. This generalization of Fibonacci numbers also leads to closed form expressions that have an algebraic and a combinatoric representation although they are substantially more complicated. The combinatoric representation for t_n for the starting value $0, 0, 1$, for example, is given by

$$t_n = \sum_{i=0}^{\lfloor n/2 \rfloor} \sum_{k=0}^{\lfloor n/3 \rfloor} \binom{n - i - 2k}{i + k} \binom{i + k}{k}$$

Such sequences easily generalize to the *k-bonacci numbers* defined by

$$(5.24) \qquad\qquad b_n = b_{n-1} + \cdots + b_{n-k}$$

with starting values given by $k-1$ zeros followed a 1. A table of these numbers up to $k=5$ follows:

k	Name	Initial Values	Rest of Sequence									
2	Fibonacci	0, 1	1	1	2	3	5	8	13	21	34	55 ...
3	Tribonacci	0, 0, 1	1	1	2	4	7	13	24	44	81	149 ...
4	Tetranacci	0, 0, 0, 1	1	1	2	4	8	15	29	56	108	208 ...
5	Pentanacci	0, 0, 0, 0, 1	1	1	2	4	8	16	31	61	120	236 ...

The defining equation for k-bonacci numbers is $x^k = x^{k-1} + \cdots + x + 1$.

It can be shown that every integer has at least one way to write it as the sum of k-bonacci integers. To illustrate this, consider the Fibonacci case where 100 can be written in terms of Fibonacci numbers as $3 + 8 + 89$, $1 + 2 + 8 + 89$, $3 + 8 + 34 + 55$ and $3 + 8 + 89$. Edouard Zeckendorf (1901–1983) showed for the $k = 2$ case that there is only one way to write such a representation that does not use adjacent Fibonacci numbers ($100 = 3 + 8 + 89$ above). It is trivial to construct a *Zeckendorf representation* for small values of n by enumeration. Assume that such a representation is possible for all integers up to n. If $n+1$ is a Fibonacci integer, then it already has such a representation (a Fibonacci number is its own representation). Therefore consider the case where $n+1$ not Fibonacci. This implies that there is a value j that satisfies

$$f_j < n+1 < f_{j+1}$$

By assumption, the value m defined by $m = n + 1 - f_j$ has a Zeckendorf representation since it is less than n. Using this representation shows that $m + f_j$ is a representation for the integer $n + 1$ that uses only Fibonacci numbers. But this representation might consist of adjacent Fibonacci numbers and thus not be a Zeckendorf representation. To show that this cannot be the case, use the general recurrence for Fibonacci integers to write

$$m + f_j < f_{j+1} = f_j + f_{j-1} \quad \Longrightarrow \quad m < f_{j-1}$$

5.5   Generalization of the Fibonacci Recurrence

Consider a generalization along different lines. Since the Fibonacci
sequence arises from the number of sequences of the numbers 1 and
2 that sum to n, it is only natural to ask what type of numbers arise
from patterns only using the numbers 1 and k. Denote the number
of such sequences by $\rho_k(n)$. Solving the general case, using a similar
argument to the derivation of equation (5.21), yields

$$\rho_k(n) = \sum_{i=0}^{\lfloor n/k \rfloor} \binom{n - (k-1)i}{i}$$

Interestingly, the recurrence equation corresponding to these values is
given by

$$g_n = g_{n-1} + g_{n-k}, \qquad n = k+1, \ldots$$

where the initial portion of the sequence consists of k, 1's, followed by
a 2. A table of these numbers is given by

k	Initial Values	Rest of Sequence											
2	1, 1, 2	3	5	8	13	21	34	55	89	144	233	377	610 ...
3	1, 1, 1, 2	3	4	6	9	13	19	28	41	60	88	129	189 ...
4	1, 1, 1, 1, 2	3	4	5	7	10	14	19	26	36	50	69	95 ...
5	1, 1, 1, 1, 1, 2	3	4	5	6	8	11	15	20	26	34	45	60 ...

Another generalization arises if the defining equation is changed so
that $x^2 = ax + 1$ for some positive integer a. An analysis similar to
that found in the beginning of the chapter shows that this creates a
Fibonacci type sequence which satisfies $h_i = ah_{i-1} + h_{i-2}$. With initial
conditions $h_0 = 0$, $h_1 = 1$, and $h_2 = a$ we find that the n'th type
Hibonacci number (coining the term) is given by

(5.25) $$h_{n,a} = \frac{(h_a^+)^n - (h_a^-)^n}{\sqrt{a^2 + 4}}$$

where

(5.26) $$h_a^+ = \frac{a + \sqrt{a^2 + 4}}{2} \quad \text{and} \quad h_a^- = \frac{a - \sqrt{a^2 + 4}}{2}$$

It is straightforward to derive equation (5.25) by mimicking the steps leading to equation (5.10). Some illustrative values of such sequences are given in the following table:

a	Hibonacci Sequence							
1	1	1	2	3	5	8	13	21	34
2	1	2	5	12	29	70	169	408	985
3	1	3	10	33	109	360	1,189	3,927	12,970
4	1	4	17	72	305	1,292	5,473	23,184	98,209
5	1	5	26	135	701	3,640	18,901	98,145	509,626
6	1	6	37	228	1,405	8,658	53,353	328,776	2,026,009
7	1	7	50	357	2,549	18,200	129,949	927,843	6,624,850

The continued fraction expansion of h_a^+ is a generalization of ϕ found in equation (5.19) and is given by[6]

$$(5.27) \qquad h_a^+ = a + \cfrac{1}{a + \cfrac{1}{a + \cfrac{1}{a + \cfrac{1}{a + \cfrac{1}{\ddots}}}}}$$

Also observe that the generalization of equation (5.18) for Hibonacci numbers is given by

$$h_a^+ = \sqrt{1 + a\sqrt{1 + a\sqrt{1 + a\sqrt{1 + \cdots}}}}$$

Clearly there are endless generalizations and results to be found along these, and other, lines.

---

[6] Derivation of equation (5.27) is found in equation (10.13) and the values in the table found on page 137.

Chapter 6
Heads I Win, Tails You Lose

> *The results concerning fluctuations in coin tossing show the widely held beliefs about the law of large numbers are fallacious. They were so amazing and so at variance with common intuition that even sophisticated colleagues doubted that coins actually misbehave as theory predicts.*

> William Feller (1906–1970)

Consider a game where two players toss a coin. If the coin lands heads up, player 1 wins a dollar. Otherwise player 2 wins a dollar. If the coin is fair, then each player has the same chance of winning. A few things about the game are obvious from the outset. Since neither player has an edge over the other, there is little chance that one of them will win a lot of money. Thus, the game should hover around break even most of the time. Additionally, each player should be ahead of the other about half of the time. Another feature of the game concerns its duration if there is an agreed stopping event. For example, suppose the game stops the first time heads is ahead of tails. Then, clearly, the game should end fairly quickly. These observations are all straightforward which suggests that coin tossing does not have much to offer in terms of mathematical results. To show this, and move on to a more interesting topic, let us quickly dispense with the mathematical analysis that establishes these obvious, intuitive, observations.

6.1 The Mathematical Model

There are many ways to create a mathematical model of how the game evolves. Let us consider one version which deals with binary numbers. If heads is thrown, then this indicated by the number 1. If tails, then

© Springer Nature Switzerland AG 2020
R. Nelson, *A Brief Journey in Discrete Mathematics*,
https://doi.org/10.1007/978-3-030-37861-5_6

by the number 0, so a game can be specified by a binary number. For example, reading left to right, the game 11011001 represents two heads, followed by a tail, then two heads, two tails, and finally ending with a head. This specification has a unique advantage since we can convert the binary integer into its decimal equivalent and use this, and the number of steps, as a way to describe a game.[1] The example just given corresponds to the 8 step game with decimal equivalent 217.

This representation gives the first result. Since a binary number of length n can represent 2^n different numbers there are 2^n different coin tossing games of length n. Another result arises from the fact that a 0 or a 1 occurs in such sequences with equal probability. This implies that all sequences of length n occur with the same probability. Thus, to calculate the probability of a particular set of games, the number of elements of this set must be determined and then multiplied by $1/2^n$. Letting $p_{k,n}$ denote the probability a coin tossing game of length n has k heads thrown. Then these comments imply that

$$(6.1) \qquad\qquad p_{k,n} = \binom{n}{k} \frac{1}{2^n}$$

To analyze this binary model, let $b_i, i = 1, \ldots, n$, be the sequence of digits ($b_i = 0$ or $b_i = 1$) of the game. The *level* of the game at step i is defined by

$$(6.2) \qquad\qquad \ell_i = 2 \sum_{k=1}^{i} b_k - i$$

In terms of the game, ℓ_i is the number of dollars player 1 is ahead (or behind if negative) of player 2 after the i'th toss. Notice that equation (6.2) shows that any permutation of the digits b_k, $k = 1, \ldots, i$, yields the same level. In particular, this means that if the digits were reversed or sorted so that heads preceded all of the tails of a sequence, or if one portion of a game was moved from the front to the back while being reversed in time, then the final level ℓ_i would not change. If one plots a game on a Cartesian coordinate axis so that tosses proceed on the x-axis and the level is plotted on the y-axis, then the game corresponds

---

[1]The number of steps also has to be given to distinguish games such as 1011 and 001011 which have the same decimal equivalence.

to a graph that steps up or steps down by one unit at each toss of the coin. With this in mind, and to avoid repetitive language, we will also refer to a game, or a portion of a game, as a *sequence* or as a *path*. For the same reason, *heads* and 1's, *tails* and 0's, will be used interchangeably when referring to the outcome of a *toss*, a *flip*, or a *step* as the game proceeds in time.

6.1.1 Games That End Even

Let $N_{m,n}$ denote the number of games of length n that end at level m. The number of heads thrown in such a game equals $(m+n)/2$. Thus, the number of such paths is given by the binomial coefficient:

$$(6.3) \qquad N_{m,n} = \binom{n}{(m+n)/2}$$

Since the belief is that most games hover around even, it makes sense to first analyze games of length $2n$ that end at level 0. The number of such games is a special case of (6.3) and, to simplify expressions, defines

$$(6.4) \qquad z_n = \binom{2n}{n}$$

To give a sense of the size of z_n consider the following upper and lower bounds. The value of z_n is the largest of all binomial coefficients in the form $\binom{2n}{k}$, $k = 0, \ldots, 2n$, and the binomial theorem shows that the sum of all such coefficients equals $2^{2n} = 4^n$. This implies that z_n must be greater than the average binomial coefficient of this type and must also be less than the total sum:

$$(6.5) \qquad \frac{4^n}{2n} < \binom{2n}{n} < 4^n$$

A *never behind* path is characterized by $\ell_{2n} = 0$ and $\ell_i \geq 0$ for $i = 1, \ldots, 2n$. This corresponds to a game that ends even but where player 1 is ahead or even throughout the entire game. From symmetry, a *never ahead* game is one where player 2 is ahead or even the entire game. In light of the opening comments, either type of game should rarely occur. A generalization of never behind or never ahead paths is to have a reference level. Thus, a never behind sequence with respect to a reference level of k is a sequence where the level never goes below level k and ends at level k. A similar definition applies to reference level of never ahead games. If a reference level is unspecified, it is assumed to be level 0.

6.1.2   Catalan Numbers

Consider a sequence ending at zero that violates the never behind condition and does so the first time at step j. At this point in the game, $\ell_{j-1} = 0$ and $\ell_j = -1$. Switch the digits from point j onward, so that a new sequence $(b_1, \ldots, b_{j-1}, \hat{b}_j, \ldots, \hat{b}_{2n})$ is created, where $\hat{b}_k = 1 - b_k$. This modified path is comprised of $n+1$, 0's, and $n-1$, 1's. There are $\binom{2n}{n+1}$ such paths. It is clear that every path that violates the never behind condition must have a first time doing so and that its modified path is unique. Thus $\binom{2n}{n+1}$ counts all paths in violation of the never behind constraint. This implies that the number of paths that are never behind, denoted by c_n, is given by

$$(6.6) \qquad c_n = \binom{2n}{n} - \binom{2n}{n+1} = \frac{1}{n+1}\binom{2n}{n}$$

These numbers, termed *Catalan* numbers, appear frequently in mathematics. They are named for Eugéne Charles Catalan (1814–1894), a Belgian mathematician who did initial work with them and who also left the world with the conjecture that 2^3 and 3^2 are the only two powers of integers that are separated by 1.[2] A list of Catalan numbers is given in the following table:

n	1	2	3	4	5	6	7	8	9	10	11
C_n	1	2	5	14	42	132	429	1,430	4,862	16,796	58,786

Another problem solved with Catalan numbers concerns the number of valid parenthesized expressions with n opening parenthesis. This solution is similar to the analysis above and follows from the observation that there cannot be a closing parenthesis without first having a

---

[2]Proved by Preda Mihăilescu (1955–) in 2002.

preceding opening parenthesis. Linking heads with opening parenthesis and tails with closing parenthesis shows that a valid parenthesized expression corresponds to a coin toss game that is never behind. There are many more models associated with Catalan numbers that have a similar flavor.

The analysis above can be levered to derive a result concerning the first time the two players are even. Let h_n denote the number of paths where level 0 is visited the first time after $2n$ tosses. To construct all such paths, consider a game that starts with a 1, is followed with $2(n-1)$ tosses that are never behind with level 1 as the reference point and then ends with a 0. Such a path visits 0 the first time at toss $2n$ and counts half of the paths that do so. By symmetry, the other half of paths are constructed by starting with a 0, followed by $2(n-1)$ tosses that are never ahead with level 1 as the reference point, and ending with a 1. The number of paths in the middle $2(n-1)$ tosses in both cases corresponds to the Catalan number, c_{n-1}. Summing these disjoint possibilities yields

$$(6.7) \qquad h_n = 2c_{n-1} = \frac{2}{n}\binom{2(n-1)}{n-1}$$

A path that ends at zero must visit zero a first time. If this occurs at toss $2k$, then from this point there are z_{n-k} paths that end at level zero. Summing all of these possibilities implies that the number of paths that visit zero at toss $2n$ equals

$$(6.8) \qquad z_n = h_1 z_{n-1} + \cdots + h_n z_0$$

Substituting equations (6.7) and (6.4) into this equation yields the identity

$$(6.9) \qquad \binom{2n}{n} = \sum_{k=1}^{n} \frac{2}{k}\binom{2(k-1)}{k-1}\binom{2(n-k)}{n-k}$$

Returning to the initial derivation of never behind paths shows that the fraction of games that satisfy this condition is given by the ratio of the number of never behind games to the number of even games

$$\frac{c_n}{z_n} = \frac{1}{n+1}$$

The fraction of even games where one player is ahead of or equal to the other player for the entire game follows from symmetry and is given by $2/(n+1)$. This is a rare event for large n. Thus, the intuition that paths frequently cross keeping both players essentially even appears to be satisfied.

6.1.3   Non-intuitive Results

There is just one problem with this conclusion—its derivation was biased by only considering paths that end even. The probability that a path ends at level m is given by equation (6.3) which is a symmetric function centered around $m = 0$. Since this corresponds to a maximum, ending even is the most likely outcome. To determine how likely this is, requires evaluating the magnitude of binomial coefficients around their midpoint. The value of a binomial coefficient can be computed using the formula (2.11) for values where k is small. For large values of n, where k is close to $n/2$, this computational method becomes infeasible. An excellent approximation in these cases is given by the *Laplace–de Moivre formula* which states that[3]

$$(6.10) \qquad \binom{n}{k} \approx 2^n \sqrt{\frac{2}{n\pi}} \, \exp\left\{-\frac{2}{n}\left(k - \frac{n}{2}\right)^2\right\}$$

This is easily programmed and is accurate for cases where k is close to the midpoint. For $n = 60$ the percent error is about 1% for $k = 20$, $.6\%$ for $k = 25$, and $.4\%$ for $k = 30$. Error percents decrease with increasing values of n.

Notice that (6.10) includes a factor of 2^n on the right-hand side of (6.10). This term conveniently cancels the denominator of $1/2^n$ when calculating the approximate probability of throwing k heads in a coin tossing game of length n, an expression given by

$$(6.11) \qquad p_{k,n} \approx \sqrt{\frac{2}{n\pi}} \, \exp\left\{-\frac{2}{n}\left(k - \frac{n}{2}\right)^2\right\}$$

---

[3]This approximation from probability theory is the only result that is not derived within the book.

This can be used to approximate the probability of being even after the $2n$'th toss

(6.12)
$$p_{n,2n} \approx \sqrt{\frac{1}{n\pi}}$$

Thus, the likelihood of being even decreases the longer the game is played. For instance, the probability that a game is even after a thousand tosses is approximately .017 which decreases to .00056 after a million tosses. This exposes a hole in the opening intuition that games hover around the break-even point. What is wrong with the intuition?

To find out, we next calculate the expected number of steps for a game to reach even when starting with one player ahead. Let e_i, $i = 0, \ldots$ be the expected number of tosses it takes until the game becomes even when the first player starts ahead by i dollars. Obviously $e_0 = 0$ and $e_i > i$ for $i \geq 1$. Consider the case where $i = 1$. There are two possibilities for the next toss. Either a tails is tossed, at which point the game is even, or a heads is tossed so that player 1 is two dollars ahead. Since each possibility occurs with equal probability the following equation can be written (the 1 in this equation counts the step):

$$e_1 = 1 + \frac{e_2}{2}$$

More generally, suppose player 1 is ahead by i dollars where $i > 1$. Then the two possibilities lead to the following equation:

$$e_i = 1 + \frac{1}{2}(e_{i-1} + e_{i+1}), \quad i = 2, \ldots$$

The general equation which summarizes these cases is given by

(6.13)
$$2e_i = \begin{cases} 2 + e_2, & i = 1 \\ 2 + e_{i-1} + e_{i+1}, & i = 2, \ldots \end{cases}$$

To solve this recurrence, first note that $e_2 = 2e_1 - 2$ and consider the case where $i = 3$. Using this and (6.13) implies

$$2e_2 = 2 + e_1 + e_3$$
$$2(2e_1 - 2) = 2 + e_1 + e_3$$
$$3e_1 - 4 = 2 + e_3$$
$$3e_1 = 6 + e_3$$

Using the same procedure for the next case yields

$$4e_1 = 12 + e_4$$

which suggests the general form

(6.14) $$ke_1 = k(k-1) + e_k, \quad k = 2, \ldots$$

To establish this, assume that it holds for all values up to k and use (6.13) to write

$$e_{k+1} = 2e_k - 2 - e_{k-1}$$
$$= 2(ke_1 - k(k-1)) - 2 - ((k-1)e_1 - (k-1)(k-2))$$
$$= (k+1)e_1 - (k+1)k$$

The pattern thus continues to the $k + 1$'th step and proves the induction. Rewriting (6.14) as

$$e_1 = k - 1 + \frac{e_k}{k}, \quad k = 2, \ldots$$

poses an immediate problem. Since $e_k/k > 1$, this equation implies that $e_1 > k$ for all k. This can only happen if e_1 is unbounded. This forces the seemingly ridiculous conclusion that a game starting with player 1 being one dollar ahead takes an infinite expected time to reach equality! Such nonsense forces us to question the validity of the analysis.

To approach the analysis in a different way, observe that equations (6.7) and (6.4) can be expressed in terms of probabilities (a *hat* is placed on a variables when converting to a probability) so that $\hat{h}_n = h_n/2^{2n}$ is the probability of first visiting zero at toss $2n$ and $\hat{z}_n = z_n/2^{2n}$ is the probability of visiting zero at step $2n$. Simple algebra reveals an interesting relationship between $\hat{h}_n$ and $\hat{z}_n$:

$$(6.15) \qquad \hat{h}_n = \frac{1}{2^{2n-1}n} \binom{2(n-1)}{n-1} = \frac{1}{2n}\hat{z}_{n-1}$$

and

$$(6.16) \quad \hat{z}_{n-1} - \hat{z}_n = \frac{1}{2^{2(n-1)}} \left(\binom{2(n-1)}{n-1} - \frac{2n(2n-1)(2n-2)!}{2^2 n^2((n-1)!)^2} \right)$$

$$= \frac{1}{2^{2(n-1)}} \binom{2(n-1)}{n-1} \left(1 - \frac{2n-1}{2n} \right)$$

$$= \frac{1}{2n}\hat{z}_{n-1}$$

Comparison of (6.15) and (6.16) shows that

$$(6.17) \qquad \hat{h}_n = \hat{z}_{n-1} - \hat{z}_n$$

A path of length $2n$ either visits zero a first time over $2n$ tosses or it never visits zero. The probability that a path does not visit zero is given by $1 - (\hat{h}_1 + \cdots + \hat{h}_n)$ which, from equation (6.17), is a telescoping sum. Since $\hat{z}_0 = 1$, this implies that

$$(6.18) \qquad \hat{z}_n = 1 - \sum_{i=1}^{n} \hat{h}_i$$

This again leads us to a non-intuitive conclusion: the probability a path never visits zero (the right-hand side of (6.18)) is the same as the probability that it visits zero at the $2n$'th toss (the left-hand side)! Expressed in terms of the game, *it is just as likely that one player leads the other for the entire game than for the players to be equal at the end of the game.* The intuition stated at the beginning of the chapter is now under attack. Before moving on to address this issue, consider an identity that arises from (6.18). Multiplying both sides of this equation by 2^{2n} with some minor algebra yields

$$(6.19) \qquad \sum_{i=1}^{n} \frac{2^{2(n-i)+1}}{n} \binom{2(i-1)}{i-1} = 2^{2n} - \binom{2n}{n}$$

In an attempt to support the initial intuition about the game, consider deriving an equation for the number of tosses that player 1 is ahead or equal to player 2 in a game of length $2n$ (player 1 is necessarily ahead an even number). Let $w_{2k,2n}$ denote the number of paths where player 1 is ahead or equal to the player 2 for $2k$ tosses in a game of length $2n$. If level 0 is not visited over the $2n$ tosses, then player 1 is either ahead the entire time or is never ahead. The probability that this occurs is given by equation (6.18) which implies two initial conditions $w_{2n,2n} = z_n$ and, by symmetry, $w_{0,2n} = z_n$. Thus the cases where $k = 0$ and $k = n$ are solved. Cases where $k = 1, \ldots, n-1$ imply that level 0 is visited at some point over the $2n$ tosses.

There are h_j paths where level zero is visited the first time on toss $2j$ and two cases to consider. Case 1 corresponds to player 1 being ahead or equal to player 2 up to the point level 0 is visited. In this case, $2j$ tosses are counted up and including visiting level 0 and $2(k-j)$ tosses are counted afterwards. Thus, the component of $w_{2k,2n}$ for case 1 equals $h_{2j}w_{2(k-j),2(n-j)}$, $j \le k$. Case 2 corresponds to player 1 being behind player 2 to the point level 0 is visited. In this case no tosses are counted over $2j$ tosses and $2k$ must be counted afterwards. The component of $w_{2k,2n}$ for case 2 thus equals $h_{2j}w_{2k,2(n-j)}$, $j \le n - k$. Each case is equally likely. Thus, summing over all j implies that (6.20)

$$w_{2k,2n} = \frac{1}{2}\left(\sum_{j=1}^{k} h_j w_{2(k-j),2(n-j)} + \sum_{j=1}^{n-k} h_j w_{2k,2(n-j)} \right), \quad k = 1, \ldots n-1$$

Like computer scientists, sometimes mathematicians need to incorporate hubris to do their work.[4] To utilize this characteristic, boldly generalize the form of the two initial cases by guessing that $w_{2k,2m} = z_k z_{m-k}$. This holds for $m = 1$ so assume it holds up to $m = n - 1$. With this guess (6.20) can be simplified to yield

(6.21) $$2w_{2k,2n} = z_{n-k} \sum_{j=1}^{k} h_j z_{k-j} + z_k \sum_{j=1}^{n-k} h_j z_{n-j-k}$$

The summations in this equation have been encountered before in equation (6.8). The first equals z_k and the second z_{n-k} which establishes the validity of the hubristic guess. Thus it is the case that

(6.22) $$w_{2k,2n} = z_k z_{n-k}$$

---

[4]Larry Wall (1954–), the author of the *Perl* programming language, listed hubris as a characteristic of a great programmer.

Moving over to probabilities with the approximation (6.12) shows that for large n

$$(6.23) \qquad \hat{w}_{2k,2n} \approx \frac{1}{\pi\sqrt{k(n-k)}}$$

Now there is a full frontal assault on intuition since both (6.22) and (6.23) are symmetric functions in k where the minimum lies in center. Intuitively, one expects the maximum of this function to be in the center which would imply that the fortunes of each player are more or less equal over the duration of the game. The implication of both equations is exactly the opposite: *it is more likely that one lucky player is ahead of the other over most of the game.*

Some illustrative values will further confute our intuition. As n increases, the approximation of (6.23) becomes more accurate. For example, setting $2n = 10,000,000$ and summing over all values k that do not result in an undefined quantity ($k = 0$ or $k = n$) yields a total probability of .9997. This is close enough to 1 to lend credence that results based on using (6.23) as a probability will approximate exact results. The object is to characterize the percent of time a game spends on one side of level 0. Since games are symmetric, these results are indifferent to whether this is player 1 or player 2. To make this characteristic precise, let α be a specified fraction and let $k(\alpha, n)$ solve

$$(6.24) \qquad \alpha = 2 \sum_{k=k(\alpha,n)}^{n-1} \frac{1}{\pi\sqrt{k(n-k)}}$$

This is not difficult to program and the results are shocking:

α	.05	.1	.15	.20	.25	.30	.35	.40	.45	.50
$k(\alpha, n)/2n$	.9985	.9940	.9865	.9759	.9625	.9461	.9271	.9053	.8811	.8546

In words, the third column shows that in 15% of games one player stays ahead of the other for 98.65% percent of the tosses. In a quarter of the games, one player leads for 96.25% percent of the time. *We are talking about some very lucky players here, not biased coins.* This is the true nature of the coin tossing game; changing sides is an infrequent occurrence. It is no wonder that intuition has failed to predict results like this.

6.2   The Correct Insight

At the beginning, the mathematical properties of coin tossing were obvious and intuitive. Analysis then abruptly transformed our facile conclusions into a quagmire of perplexity. We are at the point of puzzlement to ask the simple question: *what is really going on?* The answer is straightforward. First be assured that the above analysis is correct. The game does behave in the somewhat neurotic way outlined by the previous equations. The game is not at fault, coins do not conspire to fool us. Shakespeare says it best; in *Julius Caesar*, Cassius opines,

> *The fault, dear Brutus, is not in our stars,*
> *But in ourselves, that we are underlings.*

We need to look no further than ourselves to find the glaring fault. Common sense notions of coin tossing are just dead wrong. The root of the misunderstanding arises from the *fairness* of the game. Because there is no bias for either player to win or lose a toss, there is no preferred direction for the game. This *directionless* aspect of coin tossing subtly violates intuition.

   If there was the slightest bias in the coin to make the outcome of the game *more even*, then all of the intuitive comments made throughout the chapter would hold. For instance, suppose the coin became slightly biased towards landing up tails when player 1 was ahead. Also assume a similar bias towards landing up heads if player 2 was ahead. Under these assumptions, even if this bias was slight, then the intuition found in the opening remarks of the chapter would be satisfied. Without such a bias, there is no tendency to even out the players fortunes and the game simply drifts aimlessly along without any tendency to make things come out *fair and even*. At each toss, the game effectively starts anew from its current level. One player will be more lucky than the other. It is that simple; intuition lies on a knife edge, violated if completely balanced. It is ironic, that it is exactly the notion of *fairness* that produces outcomes that seem unfair. Having one player be ahead of the other most of the time, as we have just seen, is the game's natural state.

6.3 Particular Sequences

Let us end the chapter with some lighter fare concerning a particular sequence of heads and tails. Call a game *valid* if it does not have two consecutive 0's. How many valid games are there of length n? To calculate this, let v_n denote this quantity. Enumerating simple cases shows that $v_1 = 1$ and $v_2 = 3$. There are 5 games where $n = 3$ since only games 001, 100, and 000 violate the condition. Thus $v_3 = 5$. These examples suggest that $v_n = f_{n+2}$ where f_k is the k'th Fibonacci number.[5] The following table shows a construction method that can be used to support this supposition.

Construction Method for Calculating v_n				
$n = 2$	Append 10 to end	$n = 3$	Append 1 to end	Result for $n = 4$
		111	1111	1111
11	1110	110	1101	1110, 1101
10	1010	101	1011	1010, 1011
01	0110	011	0111	0110, 0111
		010	0101	0101

This construction illustrates that the set of valid games of length n can be obtained by forming the union of two subsets. The first appends 10 to the end of sequences in the set of valid games of length $n - 2$ and the second appends a 1 to the end of sequences in the set of valid games of length $n - 1$. It is clear that this construction produces all valid games of length n since all valid sequences must end in a 1 or in a 10. From the initial values, $v_2 = f_4$ and $v_3 = f_5$, this construction sequentially produces the Fibonacci series (since $f_i = f_{i-1} + f_{i-2}$).

To generalize, define a *valid-k* game, for $k = 2, \ldots,$ as a game where no consecutive sequence of k tails is found. Let $v_n^{(k)}$ denote the number of valid-k games. The generalization of Fibonacci numbers is that of k-bonacci numbers discussed on page 75. Instead of summing the previous two numbers to get the next value, these numbers are generated by summing the previous k numbers (see equation (5.24) along with the table on the same page). Let $f_i^{(k)}$ be the i'th k-bonacci number. Then a similar argument to the one above shows that

---

[5] See the chapter, *All that Glitters is not Gold*, especially page 67 for the definition of this sequence of numbers.

$$(6.25) \qquad\qquad\qquad v_n^{(k)} = f_{n+2}^{(k)}, \quad k = 2, \ldots$$

To illustrate the derivation of (6.25), consider Tribonacci numbers for which $k = 3$. Then games of length n consist of the union of three sets. The first appends 100 to valid-3 games of length $n-2$, the second appends 10 to valid-3 games of length $n-2$, and finally the last appends 1 to valid-3 games of length $n - 1$. This generates all $f_{n+2}^{(3)}$ possible valid-3 games. In the general case, the set of valid-k games of length n is constructed by forming the union of the set of games created by appending $\underbrace{1 \ldots 1}_{\ell-1 \; 1\text{'s}} 0$ to valid-k games of length $n - \ell$, $\ell = 1, \ldots, k$.

6.4   Conclusions

Let us sum up the chapter. A key derivation is that of the Catalan numbers which provided the mathematical framework to derive the first time a coin tossing game visits level 0. With this result, a series of results were derived that assaulted intuition. Level 0 is visited with a frequency of $1/\sqrt{n\pi}$ and is the most probable state. That being said, there is an equal probability of a game of length $2n$ ending at level 0 or never visiting level 0. Games are ultimately *unfair* in the sense that one player typically leads the other most of the time. Strangely, the expected number of steps to visit level 0 starting from any non-zero state is infinite. This fact probably goes most strongly against the experiences found in actual games.

Coin tossing is the most elementary example in the field of mathematics dealing with random walks. It takes place in one dimension, the integers of the x-axis. Higher dimensional random walks have also been extensively studied. For instance, in a random walk on the points of a plane there are four possibilities for the next step: up, down, left, or right. In three-dimensional space there are 8 possible next steps and this continues expanding as we go to higher dimensions. In continuous space, random walks are termed *diffusion processes* which have been analyzed in great detail and find applications in physics and finance. In all cases, because of the directionless property of next steps, common intuition is often violated and must be replaced with the understanding that comes through analysis.

Chapter 7
Sums of the Powers of Successive Integers

Not only could nobody but Gauss have produced it,
but it would never have occurred to any but Gauss
that such a formula was possible

Albert Einstein (1879–1955)

What happens when you sum successive powers of integers? To investigate this, define

$$(7.1) \qquad S_{k,n} = 1 + 2^k + 3^k + \cdots + n^k = \sum_{i=1}^{n} i^k, \qquad k = 0, 1, \ldots$$

An easy program generates the following table of numeric values for small k and n:

Table of Values of $S_{k,n}$									
k/n 1	2	3	4	5	6	7	8	9	10
0 1	2	3	4	5	6	7	8	9	10
1 1	3	6	10	15	21	28	36	45	55
2 1	5	14	30	55	91	140	204	285	385
3 1	9	36	100	225	441	784	1,296	2,025	3,025
4 1	17	98	354	979	2,275	4,676	8,772	15,333	25,333
5 1	33	276	1,300	4,425	12,201	29,008	61,776	120,825	220,825
6 1	65	794	4,890	20,515	67,171	184,820	446,964	978,405	1,978,405
7 1	129	2,316	18,700	96,825	376,761	1,200,304	3,297,456	8,080,425	18,080,425

The $k = 0$ case, shown above, is immediate since

$$S_{0,n} = 1^0 + 2^0 + \cdots + n^0 = n$$

© Springer Nature Switzerland AG 2020
R. Nelson, *A Brief Journey in Discrete Mathematics*,
https://doi.org/10.1007/978-3-030-37861-5_7

Supposedly, the case for $k = 1$ was assigned as a teacher's punishment
for the child prodigy, Carl Friedrich Gauss (1777–1855). Gauss was
told to sum the numbers from 1 to 100 and, instead of laboring for
an hour or two, he quickly responded 5,050 to the consternation of his
teacher. How did he do it so quickly?

The young Gauss, who later grew up to be a famous mathematician,
probably noticed that a *backwards* version of $S_{1,n}$ given by

$$S_{1,n} = n + (n-1) + \cdots + 1 = \sum_{i=1}^{n}(n+1-i)$$

could be added to the *forward* version to yield

$$2S_{1,n} = \sum_{i=1}^{n}(n+1-i) + \sum_{i=1}^{n}i = \sum_{i=1}^{n}(n+1) = n(n+1)$$

quickly giving

(7.2) $$S_{1,n} = \frac{n(n+1)}{2}$$

The precocious Gauss saw this pattern, did the numerical calculation,
and thus bypassed his teacher's punishment.

7.1 A General Equation

A key observation to make on the above approach is that by canceling
the i in the forward and backward versions of the $k = 1$ case, the
solution only required the equation for the previous, $k = 0$, case.
Following on this logic, consider another way to solve for $S_{1,n}$ which
arises by forming an equation for the next higher dimension, $k = 2$, and
having the i^2 term conveniently cancel out. To illustrate this, consider
the shifted sequence given by $(i+1)^2$. Summing this from 1 to n creates
an addition term of $(n+1)^2$ but lacks the first term when compared
to $S_{2,n}$. Thus

$$\sum_{i=1}^{n}(i+1)^2 = S_{2,n} + (n+1)^2 - 1$$

Expanding $(i + 1)^2$ and subtracting i^2 of the original sequence yields $2i + 1$ which suggests that subtracting the original sequence from its shifted version

$$S_{2,n} + (n + 1)^2 - 1 - S_{2,n} = \sum_{i=1}^{n}(i + 1)^2 - \sum_{i=1}^{n} i^2$$

$$= \sum_{i=1}^{n} 2i + 1 = 2S_{1,n} + n$$

Thus,

$$(n + 1)^2 - 1 = 2S_{1,n} + n$$

which yields

$$S_{1,n} = \frac{(n + 1)^2 - 1 - n}{2} = \frac{n(n + 1)}{2}$$

as before.

A pattern emerges which is made clear with one more example. Writing

$$(i + 1)^3 = i^3 + 3i^2 + 3i + 1$$

and using the fact that

$$\sum_{i=1}^{n}(i + 1)^3 = S_{3,n} + (n + 1)^3 - 1$$

implies that

$$S_{3,n} + (n + 1)^3 - 1 - S_{3,n} = \sum_{i=1}^{n} 3i^2 + 3i + 1 = 3S_{2,n} + 3S_{1,n} + n$$

This shows that

$$(n + 1)^3 - 1 = 3S_{2,n} + 3S_{1,n} + n$$

which, when substituting the above expression for $S_{1,n}$, solves to

$$(7.3) \qquad S_{2,n} = \frac{(n+1)^3 - (1+n) - 3n(n+1)/2}{3}$$

$$= \frac{(n+1)}{6} \left(2(n+1)^2 - 2 - 3n\right)$$

$$= \frac{(n+1)}{6} \left(2n^2 + n\right)$$

$$= \frac{n(n+1)(2n+1)}{6}$$

The pattern suggests that the solution of the k case emerges when considering the case for one dimension higher, $k+1$. An equation for $S_{k,n}$ can then be determined by arranging for the cancellation of the term, i^{k+1}. The binomial theorem implies that

$$(7.4) \quad (i+1)^{k+1} - i^{k+1} = \sum_{\ell=0}^{k+1} \binom{k+1}{\ell} i^\ell - i^{k+1} = \sum_{\ell=0}^{k} \binom{k+1}{\ell} i^\ell$$

Summation of (7.4) produces a telescoping sum on the left-hand side of the equation yielding the general equation

$$(7.5) \qquad (n+1)^{k+1} - 1 = \sum_{i=1}^{n} \sum_{\ell=0}^{k} \binom{k+1}{\ell} i^\ell$$

$$= \sum_{\ell=0}^{k} \binom{k+1}{\ell} \sum_{i=1}^{n} i^\ell$$

$$= \sum_{\ell=0}^{k} \binom{k+1}{\ell} S_{\ell,n}$$

Also observe that the binomial theorem shows that

$$(n+1)^{k+1} - 1 = \sum_{\ell=1}^{k+1} \binom{k+1}{\ell} n^\ell$$

which, after substitution into (7.5) yields

$$\sum_{\ell=0}^{k} \binom{k+1}{\ell} S_{\ell,n} = \sum_{\ell=1}^{k+1} \binom{k+1}{\ell} n^\ell$$

This can be rewritten as

(7.6) $$n + \sum_{\ell=1}^{k} \binom{k+1}{\ell} S_{\ell,n} = \sum_{\ell=1}^{k} \binom{k+1}{\ell} n^\ell + n^{k+1}$$

The right-hand side of (7.6) is a polynomial of order $k+1$ which implies that the left-hand side is also a polynomial of this order. This implies that the function $S_{\ell,n}$ can be expressed as a polynomial (previous calculations show is of order $\ell+1$) so that

$$S_{\ell,n} = a_{1,\ell}\, n + a_{2,\ell}\, n^2 + \cdots + a_{\ell+1,\ell}\, n^{\ell+1}$$

The coefficients of this polynomial can be determined by isolating powers of n. The constant coefficient above is missing since $S_{\ell,0} = 0$. Note that

$$\sum_{\ell=1}^{k} \binom{k+1}{\ell} S_{\ell,n} = \sum_{\ell=1}^{k} \binom{k+1}{\ell} \sum_{j=1}^{\ell+1} a_{j,\ell} n^j$$

$$= n \sum_{\ell=1}^{k} \binom{k+1}{\ell} a_{1,\ell} + \sum_{j=2}^{k+1} n^j \sum_{\ell=j-1}^{k} \binom{k+1}{\ell} a_{j,\ell}$$

The defining equation (7.6) can now be expressed to expose the powers of n on each side of the equation:

$$n + n \sum_{\ell=1}^{k} \binom{k+1}{\ell} a_{1,\ell} + \sum_{j=2}^{k+1} n^j \sum_{\ell=j-1}^{k} \binom{k+1}{\ell} a_{j,\ell}$$

$$= \sum_{\ell=1}^{k} \binom{k+1}{\ell} n^\ell + n^{k+1}$$

The $k+1$ equations that arise from matching powers of n^j on each side of the equation can be delineated as

Power	Equation
$j=1$	$1 + \sum_{\ell=1}^{k} \binom{k+1}{\ell} a_{1,\ell} = k+1$
$j = 2, \dots, k$	$\sum_{\ell=j-1}^{k} \binom{k+1}{\ell} a_{j,\ell} = \binom{k+1}{j}$
$j = k+1$	$(k+1) a_{k+1,k} = 1$

The last entry shows that

$$(7.7) \qquad\qquad a_{k+1,k} = 1/(k+1)$$

The remaining coefficients can be found iteratively. Two cases illustrate how this is done: For $j = 1$, we can simplify the equation found in the table

$$\sum_{\ell=1}^{k} \binom{k+1}{\ell} a_{1,\ell} = k$$

and proceed sequentially:

$$k = 1 : \binom{2}{1} a_{1,1} = 1$$

$$\implies a_{1,1} = 1/2$$

$$k = 2 : \binom{3}{1} 1/2 + \binom{3}{2} a_{1,2} = 2$$

$$\implies a_{1,2} = 1/6$$

$$k = 3 : \binom{4}{1} 1/2 + \binom{4}{2} 1/6 + \binom{4}{3} a_{1,3} = 3$$

$$\implies a_{1,3} = 0$$

$$k = 4 : \binom{5}{1} 1/2 + \binom{5}{2} 1/6 + \binom{5}{3} 0 + \binom{5}{4} a_{1,4} = 4$$

$$\implies a_{1,4} = -1/30$$

A similar procedure can be used for $j = 2$:

$$k = 1 : \binom{2}{1} a_{2,1} = 1$$

$$\implies a_{2,1} = 1/2$$

$$k = 2 : \binom{3}{1} 1/2 + \binom{3}{2} a_{2,2} = \binom{3}{2}$$

$$\implies a_{2,2} = 1/2$$

$$k = 3 : \binom{4}{1} 1/2 + \binom{4}{2} 1/2 + \binom{4}{3} a_{2,3} = \binom{4}{2}$$

$$\implies a_{2,3} = 1/4$$

$$k = 4 : \binom{5}{1} 1/2 + \binom{5}{2} 1/2 + \binom{5}{3} 1/4 + \binom{5}{4} a_{2,4} = \binom{5}{2}$$

$$\implies a_{2,4} = 0$$

The sequence of operations outlined above applies to all $j \leq k$ and is easily programmed. The results, for the first seven cases, are summarized in the following table (the column headed by d is a denominator):

$k\backslash j$	1	2	3	4	5	6	7	8	d
0	1	0	0	0	0	0	0	0	1
1	1	1	0	0	0	0	0	0	2
2	1	3	2	0	0	0	0	0	6
3	0	1	2	1	0	0	0	0	4
4	-1	0	10	15	6	0	0	0	30
5	0	-1	0	5	6	2	0	0	12
6	1	0	-7	0	21	21	6	0	42
7	0	2	0	-7	0	14	12	3	24

Table of Values of $a_{j,k}$

We have highlighted the non-zero entries in bold that are required to write the equation for $S_{5,n}$ given by

$$S_{5,n} = \frac{-n^2 + 5n^4 + 6n^5 + 2n^6}{12}$$

Using the above coefficients we can easily calculate the polynomial equation for $S_{k,n}$ but this sheds no light on the relationship between the numbers found in the table of values given at the beginning of the chapter. What is this relationship? Can one start with a simple case and generate the remainder of the numbers algorithmically?

7.1.1 Iterative Approach

If there exists such an algorithm, then it must be the case that the value of $S_{k,n}$ can be written in terms of elements that occur previous to it in the table. Typically this can be done either by a calculation of terms in the previous row or the previous column. In this way, values can be generated starting from easily calculated small values of k and n. It makes sense then to consider two different summations: one along the n axis and one along the k axis.

To follow this approach, first consider the following summation:

$$\sum_{\ell=1}^{n} S_{k,\ell} = 1 + (1 + 2^k) + (1 + 2^k + 3^k) + \cdots + (1 + 2^k + \cdots + n^k)$$

This shows that 1 appears in all n of the sums, 2^k appears in $n-1$ of them, and ℓ^k appears in $n - \ell + 1$ of the summations. Hence,

$$(7.8) \qquad\qquad S_{k,n} = nS_{k-1,n} - \sum_{i=1}^{n-1} S_{k-1,i}$$

This shows that $S_{k,n}$ can be calculated using the previous row in the table. To illustrate this, consider the case $k = 3$ and $n = 4$. The equation above yields the value of 100 that is underlined in the table:

$$100 = 4 \cdot 30 - (1 + 5 + 14)$$

Now consider the column-wise summation along the k axis. A straightforward summation like that above yields

$$\sum_{\ell=0}^{k} S_{\ell,n} = \sum_{\ell=0}^{k} \sum_{i=1}^{n} i^{\ell}$$

$$= \sum_{i=1}^{n} 1 + i^1 + i^2 + \cdots + i^k$$

$$= S_{0,n} + S_{1,n} + \cdots + S_{k,n}$$

which gets nowhere. A key insight that often proves to be useful arises by inserting a combinatorial term in the above sum, thus allowing the binomial theorem to be used to obtain a closed form equation. With this thought in mind consider the binomial transform of $S_{\ell,n}$

(7.9)
$$\sum_{\ell=0}^{k} \binom{k}{\ell} S_{\ell,n} = \sum_{\ell=0}^{k} \binom{k}{\ell} \sum_{i=1}^{n} i^{\ell}$$

$$= \sum_{i=1}^{n} \sum_{\ell=0}^{k} \binom{k}{\ell} i^{\ell}$$

$$= \sum_{i=1}^{n} (i+1)^k$$

$$= S_{k,n+1} - 1$$

This works like a charm and rewriting the above equation more directly leads to the identity

(7.10)
$$S_{k,n} = 1 + \sum_{r=0}^{k} \binom{k}{r} S_{r,n-1}$$

This equation provides a method that uses the columns to determine the value of $S_{k,n}$. Illustrating this with $k = 3$ and $n = 7$ results in the underlined value of 784 found in the table

$$784 = 1 + 6 + 3 \cdot 21 + 3 \cdot 91 + 441$$

These two methods can be used to generate all the values of $S_{k,n}$ starting from initial beginning values.

This section ends with two easily calculated identities. Forming the binomial inverse of (7.10) leads to the identity

$$(7.11) \qquad S_{k,n} = \sum_{\ell=0}^{k} \binom{k}{\ell} (-1)^{k-\ell} (S_{\ell,n+1} - 1)$$

$$= \sum_{\ell=0}^{k} \binom{k}{\ell} (-1)^{k-\ell} S_{\ell,n+1}$$

The sum of integer powers leads to

$$S_{k,n}^2 = \left(\sum_{i=1}^{n} i^k \right)^2 = S_{2k,n} + 2 \sum_{i=1}^{n-1} \sum_{j=i+1}^{n} (ij)^k$$

Thus, a closed form equation for the cross product terms contained in the summation:

$$(7.12) \qquad \sum_{i=1}^{n-1} \sum_{j=i+1}^{n} (ij)^k = \frac{S_{k,n}^2 - S_{2k,n}}{2}$$

7.2 Triangular Numbers

The values of $S_{1,n}$ given by $1, 3, 6, 10, 15, 21, 28, \ldots$ correspond to the number of items needed to fill out a triangle like a rack of billiard balls. Because of this analogy, they are called *triangular numbers*. Simplify notation and define

$$(7.13) \qquad T_n = \frac{n(n+1)}{2} = \binom{n+1}{2}, \qquad n = 0, 1, \ldots$$

Like Fibonacci numbers, triangular numbers have a multitude of interesting properties. One property was discovered by Gauss who might have had an affinity for these numbers since they allowed his school boy escape. In one of his notebooks, Gauss wrote *Eureka!* (actually his note book said, *EYPHKA*:num=$\Delta + \Delta + \Delta$) after he discovered that all numbers could be written as the sum of three triangular numbers. Hence, for any integer k, Gauss showed that it is possible to find integers m_i, $i = 1, 2, 3$, such

$$k = T_{m_1} + T_{m_2} + T_{m_3}$$

An example

$$30 = 1 + 1 + 28 = 3 + 6 + 21 = 0 + 15 + 15 = 10 + 10 + 10$$

shows that there might be multiple ways to write this sum. We say a number, n, is *represented* by a general expression if an equality for n can be found by varying the terms of that expression. Thus, Gauss's Eureka moment came when he saw that three triangular numbers represented all non-negative integers.

Endless diversion can be found in the relationships between triangular numbers. For example, straightforward algebra establishes the following identities that result in a polynomial in n:

$$(7.14) \qquad T_n + T_{n+1} = (n+1)^2$$
$$T_{n+1} - T_n = n + 1$$
$$T_{n+1}^2 - T_n^2 = (n+1)^3$$
$$8T_n + 1 = (2n+1)^2$$
$$T_{2n+1} - T_{2n} = 2n + 1$$
$$T_{2n-1} - 2T_{n-1} = n^2$$

An identity that arises from viewing triangular numbers combinatorially is given by

$$(7.15) \qquad T_{n+k} = T_n + T_k + nk$$

To prove this, consider two sets: one consisting of n elements and the other containing k elements. We are interested in calculating the number of ways to select a pair of items from these two sets, T_{n+k}. There are three mutually different ways this can be done: both elements can be selected from the set of n elements in T_n different ways, from the set of k elements in T_k different ways, and one from each set in nk different ways. Summing these yields the equation above.

By varying values of k, a variety of relations can be derived from this equation including

$$(7.16) \qquad T_{n(n+1)} = T_n + T_{n^2} + n^3 \qquad (k = n^2)$$
$$T_{n+2} = T_n + T_2 + 2n \qquad (k = 2)$$
$$T_{2n} = 2T_n + n^2 \qquad (k = n)$$
$$T_{T_n} = T_n + T_{T_{n-1}} + nT_{n-1} \qquad (k = T_{n-1})$$

By selecting $k = T_n$ in the defining equation $T_{k-1} + k = T_k$, we obtain the identity

(7.17) $$T_{T_n} = T_{T_n-1} + T_n$$

Equating the last two identities yields the tongue twisting result

(7.18) $$T_{T_n-1} = T_{T_n-1} + n T_{n-1}$$

There is a similarity between the two identities previously stated that contain an n^2 term. Stated again, these identities reveal the curious relationship

$$n^2 = T_{2n-1} - 2T_{n-1} = T_{2n} - 2T_n$$

If we assume that $T_i = 0$ for $i \le 0$, then both of these equations are special cases of a family of identities given by

(7.19) $$n^2 = T_{2n-k} + T_{k-1} - 2T_{n-k}, \quad k = 0, \ldots, n$$

A product formula, analogous to the summation identity stated above, is given by

(7.20) $$T_{nk} = T_{n-1}T_{k-1} + T_n T_k$$

Again, varying k leads to a variety of results:

(7.21) $$T_{n^2} = T_{n-1}^2 + T_n^2, \qquad\qquad (k = n)$$

$$T_{n^3} = T_{n-1}T_{n^2-1} + T_n T_{n^2}, \qquad (k = n^2)$$

$$T_{2T_n} = T_n\left(T_{n-1} + T_{n+1}\right), \qquad (k = n+1)$$

$$T_{2n} = T_1 T_{n-1} + T_2 T_n, \qquad\qquad (k = 2)$$

A triangle is a 2-dimensional object which, when raised to three dimensions, becomes a pyramid. In this case, instead of racking billiard balls, one stacks cannon balls. The sequence of values then corresponds to $1, 4, 10, 20, 35, \ldots$ which are given by values of T_{2n}. The numeric sequence can be written as $4i - 1$ and thus another identity for T_{2n} emerges into view

(7.22)
$$T_{2n} = \sum_{i=1}^{n} 4i - 1 = 4T_n - n$$

The algebra to calculate the sum of the squares of triangular numbers is a bit tricky. Linking back to the equation derived for the sums of power of integers reveals the curious relationship $S_{3,n} = T_n^2$. Using this, and the row-wise summation previously derived, shows that

(7.23)
$$\sum_{i=1}^{n} T_i^2 = \sum_{i=1}^{n} S_{3,i} = (n+1)S_{3,n} - S_{4,n}$$

Writing this in a different way leads to
(7.24)
$$\sum_{i=1}^{n} T_i^2 = \frac{1}{4} \sum_{i=1}^{n} i^2(i+1)^2 = \frac{1}{4} \sum_{i=1}^{n} i^4 + 2i^3 + i^2 = \frac{1}{4}(S_{4,n} + 2S_{3,n} + S_{2,n})$$

Equating these last two equations leads to the identity

$$2(2n+1)S_{3,n} = S_{2,n} + 5S_{4,n}$$

The general case for the summing the k'th power of triangular numbers can be written as

(7.25)
$$\sum_{i=1}^{n} T_i^k = \sum_{i=1}^{n} \frac{i^k(i+1)^k}{2^k} = \sum_{i=1}^{n} \frac{i^k}{2^k} \sum_{\ell=0}^{k} \binom{k}{\ell} i^\ell$$

$$= \frac{1}{2^k} \sum_{\ell=0}^{k} \binom{k}{\ell} \sum_{i=1}^{n} i^{k+\ell} = \frac{1}{2^k} \sum_{\ell=0}^{k} \binom{k}{\ell} S_{k+\ell,n}$$

This identity can also be recast as[1]

(7.26)
$$\sum_{i=1}^{n} i^k(i+1)^k = \sum_{\ell=0}^{k} \binom{k}{\ell} S_{k+\ell,n}$$

We can analyze the fine structure of T_i^k by using a telescoping sum to write

---

[1] This corresponds to the sum of powers of variety-2 integers, see equation (3.3).

$$(7.27) \qquad T_i^k = \sum_{j=1}^{i} T_j^k - T_{j-1}^k$$

$$= \sum_{j=1}^{i} \left(\frac{j(j+1)}{2} \right)^k - \left(\frac{(j-1)j}{2} \right)^k$$

$$= \frac{1}{2^k} \sum_{j=1}^{i} j^k \left((j+1)^k - (j-1)^k \right)$$

$$= \frac{1}{2^k} \sum_{j=1}^{i} j^k \sum_{\ell=0}^{k} \binom{k}{\ell} j^\ell \left(1 - (-1)^{k-\ell} \right)$$

$$= \frac{1}{2^k} \sum_{\ell=0}^{k} \binom{k}{\ell} \left(1 - (-1)^{k-\ell} \right) \sum_{j=1}^{i} j^{k+\ell}$$

$$= \frac{1}{2^k} \sum_{\ell=0}^{k} \binom{k}{\ell} \left(1 - (-1)^{k-\ell} \right) S_{k+\ell,i}$$

The term $1 - (-1)^{k-\ell}$ above equals 0 if $k - \ell$ is even and equals 2 if $k - \ell$ is odd. In light of this parity, let $\mathcal{I}_k$ be the set of odd (even, respectively) integers less than k if k is an even (respectively, odd) integer. Then, summing the above equation yields

$$(7.28) \qquad \sum_{i=1}^{n} T_i^k = \frac{1}{2^{k-1}} \sum_{i=1}^{n} \sum_{\ell \in \mathcal{I}_k} \binom{k}{\ell} S_{k+\ell,i}$$

$$= \frac{1}{2^{k-1}} \sum_{\ell \in \mathcal{I}_k} \binom{k}{\ell} \sum_{i=1}^{n} S_{k+\ell,i}$$

$$= \frac{1}{2^{k-1}} \sum_{\ell \in \mathcal{I}_k} \binom{k}{\ell} \left((n+1) S_{k+\ell,n} - S_{k+\ell+1,n} \right)$$

The last identity of this section displays a curious symmetry

$$(7.29) \qquad n^2 T_k + n T_{k-1} = k^2 T_n + k T_{n-1}$$

7.3 Cauchy's Theorem

As stated before, Gauss showed that all numbers could be written as the sum of three triangular numbers. This generalizes to squares with a theorem by Joseph-Louis Lagrange (1736–1813) which shows that four squares summed together are sufficient to represent all non-negative integers (like triangular numbers we consider 0 to be a member of this set). A generalization of these results is best stated in terms of polygonal numbers.

The n'th number from the set of k polygonal numbers is given by

$$P_{k,n} = \frac{(k-2)n^2 - (k-4)n}{2}$$

Triangular numbers correspond to $k = 3$ in the above equation and square numbers to $k = 4$. Like racking billiard balls, these numbers arise when you rack k-gons. Polygonal numbers can be written multiple ways in terms of triangular numbers after some algebraic manipulation

$$(7.30) \quad P_{k,n} = (k-2)T_{n-1} + n = (k-3)T_{n-1} + T_n = (k-4)T_{n-1} + n^2$$

Some conjectures are so pure and beautiful that one feels that something would be wrong with the world if they were not true. We are facing one example here because the natural generalization of the two cases above is: *all integers can be written as the sum of k polygonal numbers of order k*. It turns out that this is not just a lofty conjecture that implies the perfection of mathematics, it is actually a theorem that was first proved by Augustin-Louis Cauchy (1789–1857).

It might seem that there are good reasons to not expect Cauchy's result since the distance between successive polygonal numbers increases with k and n

$$P_{k,n} - P_{k,n-1} = n(k-2) + 3 - k$$

For example, the $k = 8$ sequence corresponds to an *octagon* whose first few values are

$$0,\ 1,\ 8,\ 21,\ 40,\ 96,\ 133,\ 176,\ 225,\ 280$$

whereas for squares we have

$$0,\ 1,\ 4,\ 9,\ 16,\ 25,\ 36,\ 49,\ 64,\ 81$$

The increased spacing of the octagonal case over the square case is offset by the need to have eight, rather than four, values added together so that all integers can be represented. This is a deep and beautiful result.

It is hard to resist answering one further question that arises: *are some k* polygonal *numbers also triangular numbers?* Recall the identity $n^2 = T_{2n-1} - 2T_{n-1}$ of (7.14). Using this equation and the identity $P_{k,n} = (k-4)T_{n-1} + n^2$ shows that these two equations share the common terms of n^2 and T_{n-1}. Equating them

$$n^2 = T_{2n-1} - 2T_{n-1} = P_{k,n} - (k-4)T_{n-1}$$

thus uncovers that $k = 6$ satisfies the equation and thus hexagonal k-gons are also triangular,

$$P_{6,n} = T_{2n-1}$$

What other forms of equations can be used to represent sets of numbers? How about the number of cubes needed to represent all numbers or, for that matter, the number of k powers that are required. Edward Waring (1736–1798) posed this question by defining a function $g(k)$ to be the minimum number of k powers that were required to represent all of the integers. Lagrange's four square theorem showed that $g(2) = 4$. Currently we only know the values $g(3) = 9$, $g(4) = 19$, $g(5) = 37$, and $g(6) = 73$. Bounds and properties of g have been extensively studied but, as of yet, the functional form of g remains unknown.

Chapter 8
As Simple as $2 + 2 = 1$

Freedom, is the freedom to say two plus two make four.
If that is granted, all else follows.

George Orwell (1903–1950)

Orwell was not speaking about mathematics in the quote above from his book *1984*. Rather, he was commenting on how totalitarian governments attempt to define, and impose, their own notion of reality on the public. Speaking mathematically, it is as clear as the back of your hand that $2 + 2 = 1$ and $1 + 2 = 0$. That is, if you belong to a three fingered species. We have grown so used to the ten fingers on our hands, that we forget that there is nothing special about base 10. Since the invention of the *number* 0 by Indian mathematicians of the fifth century, this means that all of our numbers are composed of the digits 0 through 9. To three fingered species this means that their number system uses the digits 0 through 2 so that 3 wraps around to 0 and 4 to 1. Thus $2 + 2 = 1$ and $1 + 2 = 0$ in base 3. Orwell's above statement is thus valid for all bases 5 and larger unless, of course as he alludes, the totalitarian regime in power says otherwise.

8.1 Modular Arithmetic

Often the remainder of a number after division is the only characteristic that is necessary to establish a mathematical property—the magnitude of the integer is not relevant. For example, all primes greater than 2 are odd independent of their magnitude. In base 2 looking at integers in this way essentially splits them into 2 disjoint sets. The first set contains the even integers, $\{\ldots, -4, -2, 0, 2, 4, \ldots\}$

© Springer Nature Switzerland AG 2020
R. Nelson, *A Brief Journey in Discrete Mathematics*,
https://doi.org/10.1007/978-3-030-37861-5_8

and the second the odds, $\{\ldots, -5, -3, -1, 1, 3, 5, \ldots\}$. These two groups arise by skipping the integers by 2 (up and down) starting from a point determined by the remainder when an integer is divided by 2.

In base n integers are split into n disjoint sets depending on their remainder when divided by n (the possible remainders are 0 through $n - 1$). Similar to odd even parity, these sets occur by skipping by n steps. For example, the set corresponding to a remainder of k is given by

$$\{\ldots, -3n + k, -2n + k, -n + k, k, n + k, 2n + k, \ldots\},$$
$$k = 0, \ldots, n - 1$$

Mathematically, one uses modular arithmetic when the only characteristic necessary to establish a property is the parity of the number with respect to some base. Equality in such a system is customarily written as

$$b \equiv \beta \ (\text{mod } n)$$

which means that b and β leave the same remainder when divided by *modulus* n. This equation represents a *congruence relation* between b and β. A more concise notation used in this book when dealing with *modulo arithmetic* is written as

(8.1) $$b \equiv_n \beta$$

As examples, the equations $38 \equiv_5 53$, $-2 \equiv_5 3$, and $-47 \equiv_5 -222$ are all valid since 38, 53, -2, -47, and -222 leave a remainder of 3 when divided by 5. Since $an + k \equiv_n bn + k$ for any integers a and b it is customary to write the right-hand side of a modulo equation by setting $b = 0$ or $b = -1$. Thus the equation $38 \equiv_5 53$ would typically be written as $38 \equiv_5 3$ or $38 \equiv_5 -2$.

The equation $b \equiv_n \beta$ is equivalent to the fact that $b - \beta$ is evenly divisible by n. In essence, both statements are a restatement of the equation $(b - \beta) \equiv_n 0$. In terms of the sets that we mentioned above, the equation means that b and β are in the same set. Negative numbers in modulo arithmetic can be viewed in terms of positive complements through the following equation:

(8.2) $$n - b \equiv_n -b, \quad b = 0, \ldots, n - 1$$

For example, $9 \equiv_{10} -1$. In everyday life, clocks form a natural modulo system of order 12 provided that the hours are relabeled 0 through 11.

It is clear that the congruence relation is *reflexive* ($b \equiv_n b$), *symmetric* ($b \equiv_n \beta$ also means $\beta \equiv_n b$), and *transitive* ($b \equiv_n \beta$ and $\beta \equiv_n \gamma$ imply $b \equiv_n \gamma$). Other properties of modulo arithmetic include (ℓ is assumed to be an integer)

$$\ell + b \equiv_n \ell + \beta \qquad \text{addition property}$$

$$\ell b \equiv_n \ell \beta \qquad \text{multiplication property}$$

$$b^\ell \equiv_n \beta^\ell \quad (\ell > 0) \quad \text{power property}$$

One way to establish the last property is to use (A.7) to write

$$b^\ell - \beta^\ell = (b - \beta) \sum_{i=0}^{\ell-1} b^i \beta^{\ell-i-1}$$

The divisibility of $b^\ell - \beta^\ell$ by n then follows from the fact that $b - \beta$ is divisible by n. Notice that the combination of all three properties listed above imply that if p is a polynomial with integer coefficients and $b \equiv_n \beta$, then $p(b) \equiv_n p(\beta)$.

To state properties having combinations of modular terms, assume that $a \equiv_n \alpha$ and $b \equiv_n \beta$. Then

(8.3)
$$a \pm b \equiv_n \alpha \pm \beta$$

$$ab \equiv_n \alpha \beta$$

To establish the last equation, note that the quantity $ab - \alpha\beta = \alpha(b - \beta)$ is divisible by n and thus $\alpha b \equiv_n \alpha \beta$. Similarly $ab - \alpha b = b(a - \alpha)$ is divisible by n and thus $ab \equiv_n \alpha b$. These two equations, along with symmetry and transitivity, yield $ab \equiv_n \alpha \beta$. This relationship provides another way to establish the power property in the first list, $b^\ell \equiv_n \beta^\ell$. To see this, set $a = b$ and $\alpha = \beta$ and repeat $\ell - 1$ times.

The following two properties can be used to cancel ℓ in the following equations:

$$\ell + b \equiv_n \ell + \beta \quad \Longrightarrow \quad b \equiv_n \beta$$

$$\ell b \equiv_n \ell \beta \quad \Longrightarrow \quad b \equiv_n \beta$$

provided that n and ℓ are coprime

In this last equation, n and ℓ must not have any common factors for the relationship to hold in general. To show this, write $\ell b - \ell \beta = \ell(b - \beta)$. If ℓ is divisible by n, then there is no necessity for $b - \beta$ to also be divisible by n. Thus ℓ cannot be cancelled from the equation and still guarantee the equality. If ℓ and n are co-prime, however, then the only way $\ell(b - \beta)$ is divisible by n is for $b - \beta$ to be divisible by n. Thus $b \equiv_n \beta$.

8.2   Fermat's Little Theorem

At this point it makes sense to ask—what use is such a concept? Many applications seem like tricky test questions. To illustrate one example: what is the remainder when 19^{317} is divided by 3? To determine the answer, note that $19 \equiv_3 1$ since $19 = 3 \cdot 6 + 1$. The power property above then yields the answer: $19^{317} \equiv_3 1$.

For another example, let ℓ be an integer with digits d_i, $i = 0, \ldots, m$ where the i'th digit corresponds to the i'th power of 10. This corresponds to the polynomial

$$\ell = d_0 + d_1 10^1 + \cdots + d_m 10^m$$

Suppose that $\ell \equiv_3 0$. Then is it possible to say anything about the digits comprising ℓ? To answer this, note that $10 \equiv_3 1$ and thus $10^i \equiv_3 1$. Using the multiplication property shows that $d_i 10^i \equiv_3 d_i$ and thus $\ell \equiv_3 d_0 + d_1 + \cdots + d_m$. These observations imply that the sum of the digits of ℓ must be divisible by 3 for ℓ to be divisible by 3. A straightforward generalization concerns base n integers written as

$$\ell = d_0 + d_1 n^1 + \cdots + d_m n^m$$

A similar argument shows that $\ell \equiv_{n-1} d_0 + \cdots + d_m$ (since $n \equiv_{n-1} 1$). Thus, octal integers are divisible by 7 if the sum of their digits is divisible by 7.

As another example, a number is divisible by 11 if the alternating ($\pm$) sum of its digits is divisible by 11 (this arises from the fact that $10 \equiv_{11} -1$). There are a wealth of results along these lines.

Modulo arithmetic also leads to many basic results of number theory. For example, assume that p is a prime number and consider the binomial expansion

$$(a+b)^p = \sum_{i=0}^{p} \binom{p}{i} a^i b^{p-i}$$

$$= a^p + b^p + \sum_{i=1}^{p-1} \binom{p}{i} a^i b^{p-i}$$

In this expression, the binomial coefficient in the summation is divisible by p since p is contained in the numerator[1] but not the denominator. Hence

$$\binom{p}{i} \equiv_p 0, \quad i = 1, \ldots, n-1$$

which implies that

(8.4) $$(a+b)^p \equiv_p a^p + b^p$$

A result that follows from this is due to Pierre de Fermat (1607–1665) which is aptly called *Fermat's Little Theorem*. It states that

(8.5) $$x^p \equiv_p x$$

for x integer and p prime. To prove it, observe that the claim clearly holds for $x = 0$. Thus, assume it holds up to a value of $x = a$. Using (8.4) we can write

$$(a+1)^p \equiv_p a^p + 1^p \equiv_p a + 1$$

where the last step follows from the induction assumption. This equation is simply a restatement of (8.5) for the next highest integer and establishes the result.

Fermat's little theorem can be used as a test to determine if an integer n is prime. For example, suppose for some x that $x^n \not\equiv_n x$. Then the theorem implies that n is not prime. No statement can be made; however, if Fermat's equation holds since it can do so for composite n. In fact, there are composite integers n that satisfy Fermat's equation for every value x that is relatively prime to n.

---

[1] The numerator equals $p(p-1)!$.

Such *Carmichael numbers* pose a formidable test to Fermat since they present the appearance of being prime. There are an infinite number of such Carmichael numbers masquerading as primes, at least through the eyes of Fermat's test. The smallest such number is 561.

8.3   Lagrange's Theorem

Another result from modular arithmetic is due to Lagrange and deals with the roots of a polynomial modulo a prime. A polynomial $f(x) = a_0 + a_1 x + \cdots + a_m x^m$ has degree k modulo n if a_k is the highest coefficient that is not divisible by n. The theorem states that the number of roots of a polynomial modulo p, where p is prime, cannot exceed its degree. The proof of this proceeds by induction starting with a degree 1 polynomial where the result is obvious. Assume the proposition holds up to degree $n-1$. Suppose polynomial f has degree n modulo p. If f does not have a root, then there is nothing to prove. Therefore, assume that a root b exists so that $f(b) \equiv_p 0$. Write

$$f(x) - f(b) = \sum_{i=1}^{n} a_i (x^i - b^i)$$

$$= (x - b) \sum_{i=1}^{n} a_i \sum_{j=0}^{i-1} x^j b^{i-j-1} \qquad \text{from equation (A.7)}$$

$$= (x - b) \sum_{j=0}^{n-1} x^j \sum_{i=j+1}^{n} a_i b^{i-j-1}$$

$$= (x - b) \sum_{j=0}^{n-1} c_j x^j$$

where c_j is defined as

$$c_j = \sum_{i=j+1}^{n} a_i b^{i-j-1}$$

Thus $f(x)$ can be written as

$$f(x) = f(b) + (x - b) \sum_{j=0}^{n-1} c_j x^j \equiv_p (x - b) \sum_{j=0}^{n-1} c_j x^j$$

By the induction hypothesis, the polynomial $\sum_{j=0}^{n-1} c_j x^j$ can have at most $n - 1$ roots. This, along with the assumption that b is a root, implies there can be at most n roots to $f(x)$ modulo p and thus establishes the result.

8.4 Wilson's Theorem

The theorems of Fermat and Lagrange just discussed can be used to extract a deep result. Fermat's result (8.5) can be rewritten as $x^{p-1} - 1 \equiv_p 0$ which shows that there are $p - 1$ roots modulo p with the values $1, 2, \ldots, p-1$. Consider the polynomial (see (3.1) for the falling factorial notation)

$$(x - 1)^{\underline{(p-1)}} = (x - 1)(x - 2) \cdots (x - (p - 1))$$

This clearly also has roots $1, 2, \ldots, p - 1$ modulo p. Using the result (3.13) we can write

$$(x - 1)^{\underline{(p-1)}} = \sum_{i=1}^{p-1} (-1)^{p-i-1} \begin{bmatrix} p - 1 \\ i \end{bmatrix} (x - 1)^i$$

$$= \sum_{i=1}^{p-1} (-1)^{p-i-1} \begin{bmatrix} p - 1 \\ i \end{bmatrix} \sum_{j=0}^{i} \binom{i}{j} (-1)^{i-j} x^i$$

$$= \sum_{i=1}^{p-1} (-1)^{p-1} \begin{bmatrix} p - 1 \\ i \end{bmatrix}$$

$$+ \sum_{j=1}^{p-1} x^j \sum_{i=j}^{p-1} (-1)^{p-j-1} \binom{i}{j} \begin{bmatrix} p - 1 \\ i \end{bmatrix}$$

$$= (-1)^{p-1} \begin{bmatrix} p - 1 \\ p - 1 \end{bmatrix} + \sum_{j=1}^{p-1} e_{j,p-1} x^j$$

where we have defined

$$e_{j,p-1} = \sum_{i=j}^{p-1} (-1)^{p-j-1} \binom{i}{j} \begin{bmatrix} p - 1 \\ i \end{bmatrix}, \quad j = 0, \ldots, p - 1$$

If p is a prime greater than 2, then we can use (3.5), and the observation
that $c_{p-1,p-1} = 1$, to simplify the above expression:

$$(x-1)^{\underline{(p-1)}} = (p-1)! + x^{p-1} + \sum_{j=1}^{p-2} e_{j,p-1} x^j$$

Our next step is to consider the polynomial defined by subtracting
Fermat's equation from the falling factorial

$$(8.6) \qquad f(x) = (x-1)^{\underline{(p-1)}} - \left(x^{p-1} - 1\right)$$

$$= (p-1)! + 1 + \sum_{j=1}^{p-2} e_{j,p-1} x^j$$

$$= \sum_{j=0}^{p-2} e_{j,p-1} x^j$$

In the last equation we have extended the definition of the coefficients
to include the constant term

$$(8.7) \qquad\qquad e_{0,p-1} = (p-1)! + 1$$

From its definition, it is clear that f has the same roots as the two
functions that define it. Thus it has $p-1$ roots modulo p. But this is
impossible according to Lagrange's theorem since f has of degree $p-2$
modulo p which only allows $p-2$ roots. Hence, f must be identically
equal to 0 modulo p which implies that all of its coefficients must equal
0 modulo p:

$$e_{j,p-1} \equiv_p 0, \quad j = 0, \ldots, p-2$$

This establishes the following set of identities:

$$(8.8) \qquad\qquad (p-1)! + 1 \equiv_p 0$$

$$\sum_{i=j}^{p-1} (-1)^{p-j-1} \binom{i}{j} \begin{bmatrix} p-1 \\ i \end{bmatrix} \equiv_p 0, \quad j = 1, \ldots, p-2$$

These ruminations lead us to the deep result mentioned above. Wilson's theorem named after John Wilson (1741–1793) states that the equation

$$(8.9) \qquad (n-1)! + 1 \equiv_n 0$$

is satisfied if and only if n is a prime number. The if portion is simply the first identity above (8.8) that deals with the constant coefficient of f. To prove the only if portion of Wilson's theorem, assume that n is composite and also satisfies $(n-1)! + 1 \equiv_n 0$. Then it must be the case that n is divisible by an integer $k < n$. But $(n-1)!$ necessarily contains k and thus $(n-1)! + 1 \not\equiv_k 0$ for any $k < n$. This violates the assumption that n is composite.

Wilson's theorem provides another method to test if a number is prime—simply see if $(n-1)! + 1 \equiv_n 0$. Like Fermat's little theorem, it also is a poor test since factorials increase rapidly. Wilson's theorem does provide entertainment by producing a wealth of parlor tricks. For example, from the theorem we know that $72! \equiv_{73} -1$ since 73 is prime. Thus, since $72 \equiv_{73} -1$, we can use (8.3) to conclude that $71! \equiv_{73} 1$. Following this example, we can write the general equation $(p-2)! \equiv_p 1$ for p prime.

8.5 Cryptography

A less flippant application of modular arithmetic than numeric divisibility challenges is a procedure that is used millions of times a day on the Internet. It is called *public key cryptography*. The objective of cryptography is to create a secure communications channel between two participants such that an eavesdropper cannot decode their communications. To explain this procedure, suppose that A wants to send a secure message to B and that C is listening to the communication. Both A and B share a large prime number p and a base number n. These can also be known by C. Let e_a be an integer only known by A and similarly let e_b be an integer only known by B. Participant A computes the value

$$v_a \equiv_p n^{e_a}$$

and sends it on the channel to B. Likewise, B computes

$$v_b \equiv_p n^{e_b}$$

and sends it to A. Note that C can listen to these communications; they take place on an insecure channel.

Now the magic starts. Participant A takes the communication it received from B and computes

$$w_a \equiv_p v_b^{e_a}$$

This computed value cannot be determined by C because e_a is only known to A. Likewise, B computes

$$w_b \equiv_p v_a^{e_b}$$

which also cannot be computed by C. But A and B now share the same value because $w = w_a = w_b$. This follows from the power property of modular arithmetic since

$$v_b \equiv_p n^{e_b} \quad \Longrightarrow \quad (v_b)^{e_a} \equiv_p (n^{e_b})^{e_a}$$

and

$$v_a \equiv_p n^{e_a} \quad \Longrightarrow \quad (v_a)^{e_b} \equiv_p (n^{e_a})^{e_b}$$

The value of w can now be used as a key for a cryptographic scheme for the duration of the communication and C cannot *easily* break the code because everything is calculated modulo a large prime. I need the word *easily* in this last statement because if there was a fast way to calculate the value of e_a or e_b from w (remember p, n, v_a, and v_b are all assumed to be known by C), then C could break the code. Solving for such values is called the *discrete logarithm problem* which is currently computationally intractable for large p.

Chapter 9
Hidden in Plain Sight

> *Chebyshev said it, and I'll say it again,*
> *There's always a prime between n and 2n*
>
> Nathan Fine (1916–1994)

Take any number and keep finding factors of that number that cannot be factored themselves. For example, $84 = 2 \cdot 2 \cdot 3 \cdot 7$, $455 = 5 \cdot 7 \cdot 13$ or $897 = 3 \cdot 13 \cdot 23$. These examples show that a number can be written as the product of prime numbers.[1] This is called a *prime factorization*. A separate argument, that we will shortly get to, shows that this factorization is unique. This result has far reaching consequences and is called the *Fundamental Theorem of Arithmetic*. This theorem shows that primes are the DNA of the number system. Essentially all of the results of number theory are theorems of the primes, the topic of this chapter.

9.1 Properties of Prime Numbers

Do primes ever end? To address this question, let $p_1, p_2, \ldots, p_n$ be a list of successive primes that end with an assumed maximal prime p_n. Consider a *primorial number*, denoted by the somewhat strange notation $p_n\#$, that corresponds to the product of the first n consecutive primes

$$(9.1) \qquad p_n\# = p_1 \cdot p_2 \cdots p_{n-1} \cdot p_n$$

---

[1] Just keep dividing until it is not possible to continue without having a remainder.

© Springer Nature Switzerland AG 2020
R. Nelson, *A Brief Journey in Discrete Mathematics*,
https://doi.org/10.1007/978-3-030-37861-5_9

This is not a prime number but how about $p_n\# + 1$? If $p_n\# + 1$ is divided by p_j:

$$\frac{p_n\# + 1}{p_j} = p_1 \cdot p_2 \cdots p_{j-1} \cdot p_{j+1} \cdots p_{n-1} \cdot p_n + \frac{1}{p_j}$$

then the result is a whole number with a remainder. But if no previous prime divides $p_n\# + 1$, then it must be prime and it is clearly larger than p_n, the presumed largest prime. This contradicts the assumption that there is a maximal prime. This clever argument was first put forth sometime around 300 BC by Euclid of Alexandria, the father of *Euclidean geometry*. There are literally dozens of proofs that the primes go on infinitely and we will see a couple more in this chapter.

How does one generate a list of prime numbers? Let us discuss one way proposed by another Greek mathematician, Eratosthenes (276–194 BC), who developed a technique sometime around 200 BC. It is a simple idea called a *sieve*. For a variety of reasons the integer 1 is not considered to be prime.[2] A sieve starts by writing all the integers starting from 2

$$2, 3, 4, 5, 6, 7, 8, 9, 10, 11, 12, 13, 14, 15, 16, 17, 18, 19, 20, 21, 22, 23, \ldots$$

The first number in the list, 2, is the first prime and this means that all subsequent multiples of 2 can be crossed out as candidate primes;

$$2, 3, \not{4}, 5, \not{6}, 7, \not{8}, 9, \not{10}, 11, \not{12}, 13, \not{14}, 15, \not{16}, 17, \not{18}, 19, \not{20}, 21, \not{22}, 23, \ldots$$

This leaves the next prime, 3. Repeating this process, cross out all multiples of 3 yielding

$$2, 3, \not{4}, 5, \not{6}, 7, \not{8}, \not{9}, \not{10}, 11, \not{12}, 13, \not{14}, \not{15}, \not{16}, 17, \not{18}, 19, \not{20}, \not{21}, \not{22}, 23, \ldots$$

Notice that 6 is already crossed out because it was divisible by 2 and would again be crossed out because it is also divisible by 3. The first remaining number is the next prime, 5, and the process continues by crossing out multiples of 5

$$2, 3, \not{4}, 5, \not{6}, 7, \not{8}, \not{9}, \not{10}, 11, \not{12}, 13, \not{14}, \not{15}, \not{16}, 17, \not{18}, 19, \not{20}, \not{21}, \not{22}, 23, \ldots$$

This leaves 7 as the next prime. The algorithm continues on from here.

---

[2] Essentially it makes too many trivial exceptions in theorems in number theory.

Eventually the only numbers left not crossed are the set of primes. This sieve shows how primes emerge as the essential components of the integers. There are other methods to generate primes and long lists of primes can be found on the Internet. Its maddening, however, that there is no equation for the n'th prime. No one who has walked the earth, and even perhaps who will ever walk the earth, knows the 17 quadragintillion'th prime.

We next return to the fundamental theorem of arithmetic and prove that the representation of an integer as a product of primes is unique. For small values of n this is easily established so assume the first time uniqueness does not hold is at integer m. It is clear that m cannot be prime. Assume that composite m has two different prime factorizations. In these factorizations, order the primes in increasing value so that p is the smallest prime in the first factorization and q is the smallest in the second. Let the remaining portion of the factorizations of m be denoted by P and Q, respectively. Thus we can write

$$m = pP = qQ$$

Note that P is composed of primes at least as large as p and, similarly, Q consists of primes at least as large as q. If $p = q$, then we reach a contradiction since $P = m/p = m/q = Q$ is an integer smaller than m which is assumed to have a unique factorization. Since p and q differ, one has to be greater so assume that $p > q$. Note, from the two factorizations above, m is divisible by both p and q. It clearly is divisible by p but what about q? If q divides m, then it must divide either p or P. But this is impossible since both of these terms consist of products of primes that are strictly larger than q. Thus we reach a contradiction—prime factorizations are unique.

From now we will talk about integers in terms of their prime factorization. Let $\omega_i(n)$ be the exponent of the i'th prime, p_i, in the representation of the integer n, thus

$$n = 2^{\omega_1(n)} \cdot 3^{\omega_2(n)} \cdot 5^{\omega_3(n)} \cdot 7^{\omega_4(n)} \cdots p_i^{\omega_i(n)} \cdots = \prod_{i=1}^{\infty} p_i^{\omega_i(n)}$$

To multiply two numbers using this representation one simply adds exponents

$$n \cdot m = \prod_{i=1}^{\infty} p_i^{\omega_i(n)+\omega_i(m)}$$

More conveniently we can write n as the infinite vector of values $\omega(n) = (\omega_1(n),\ \omega_2(n),\ \ldots)$ where the j'th place corresponds to the exponent of p_j. Thus $n \cdot m$ is represented as the vector addition: $\omega(n \cdot m) = (\omega_1(n) + \omega_1(m),\ \omega_2(n) + \omega_2(m),\ \ldots)$. As an example,

$$(9.2) \qquad\qquad 198 = (1,2,0,0,1,0,\ldots) = 2 \cdot 3^2 \cdot 11$$

Note that $w_i(n/m) = w_i(n) - w_i(m)$ (provided n is divisible by m) and $w_i(n^k) = k w_i(n)$ for integer k.

9.1.1 Properties of Integer Divisors

This notation allows us to specify the total number of divisors as well as the sum of all the divisors with simple arithmetic expressions. Define $\sigma_k(n)$ be the

$$(9.3) \qquad\qquad \sigma_k(n) = \sum_{d \mid n} d^k$$

where $d \mid n$ means that the summation occur over all possible integer values of d that evenly divide n. Thus, $\sigma_0(n)$ is the number of total divisors of n and $\sigma_1(n)$ is the sum of the total divisors. For the example (9.2) above there are 12 divisors given by

$$\{1,2,3,11,2 \cdot 3, 2 \cdot 11, 3^2, 3 \cdot 11, 2 \cdot 3^2, 2 \cdot 3 \cdot 11, 3^2 \cdot 11, 2 \cdot 3^2 \cdot 11\}$$

To calculate an expression for $\sigma_0(n)$, concentrate on the i'th prime which has an exponent of $\omega_i(n)$. The total number of possible divisors due to this prime is given by

$$p_i^0,\ p_i^1, \ldots, p_i^{\omega_i(n)}$$

leading to a total of $1 + \omega_i(n)$ possibilities. Since this is true for all i we have

$$(9.4) \qquad \sigma_0(n) = (1 + \omega_1(n))(1 + \omega_2(n)) \cdots = \prod_{i=1}^{\infty} (1 + \omega_i(n))$$

Note the special case for primes raised to a power:

(9.5)
$$\sigma_0(p_i^k) = k + 1$$

The value sum of the divisors for the example can be expressed by

$$\sigma_1(198) = (2^0 + 2^1)(3^0 + 3^1 + 3^2)(11^0 + 11^1)$$

To explain this, note that expanding this multiplication into its individual factors corresponds to summing all possible products of the form $2^{k_1}3^{k_2}11^{k_5}$, where $k_1 = 0, 1$, $k_2 = 1, 2, 3$, and $k_5 = 1, 2$. Following this pattern we can write the sum of all the divisors for n as the product of the sum of all the divisors for the i'th prime

$$1 + p_i^1 + p_i^2 + \cdots + p_i^{\omega_i(n)} = \frac{p_i^{\omega_i(n)+1} - 1}{p_i - 1}$$

Thus $\sigma_1(n)$ is given by

(9.6)
$$\sigma_1(n) = \prod_{i=1}^{\infty} \frac{p_i^{\omega_i(n)+1} - 1}{p_i - 1}$$

For the example above, we obtain $\sigma_1(198) = 468$. Note the special case for primes raised to a power:

(9.7)
$$\sigma_1(p_i^k) = 1 + p_i^1 + \cdots + p_i^k = \frac{p_i^{k+1} - 1}{p_i - 1}$$

The prime factorization of $n! = n(n-1)\cdots 2 \cdot 1$ can be obtained from the factorizations of all the integers less than or equal to n. For prime p_i, let $\Omega_i(n) = \omega_i(n!)$ which is given by

$$\Omega_i(n) = \sum_{k=2}^{n} \omega_i(k)$$

A modification of a sieve argument allows us to calculate a closed form expression for $\Omega_i(n)$. To motivate this argument, consider the prime factorization of 10! and concentrate on the value of the exponent of 2 in that factorization. Every step of size 2^i, $i = 1, \ldots$ corresponds to a multiplication by 2 which has to be counted in the final value of the exponent. For 10! we have factors associated with 2^1, arising from $2, 4, 6, 8, 10$, factors associated with 2^2, from 4 and 8, and associated with 2^3, from 8. Counting all of these shows that the final exponent of 2 equals 8. Mathematically, we can write the contribution to the final exponent for factors associated with 2^i by the integer portion of $n/2^i$. Generalization of this argument shows that

$$(9.8) \qquad \Omega_i(n) = \sum_{\ell : p_i^\ell \leq n} \left\lfloor \frac{n}{p_i^\ell} \right\rfloor$$

The prime factorization can therefore be written as

$$(9.9) \qquad n! = \prod_{i=1}^{\infty} p_i^{\Omega_i(n)}$$

From this equation it is clear that $n!$ cannot be prime (indeed all factorial numbers are even), but $n! \pm 1$ could be. Such primes are called *factorial primes*. Less than a hundred of these primes have been discovered.

Recall the definition of a primorial number defined in equation (9.1), values of which are given by

n	1	2	3	4	5	6	7	8	9
p_n	2	3	5	7	11	13	17	19	23
$p_n\#$	2	6	30	210	2,310	30,030	510,510	9,699,690	223,092,870

9.2   The Prime Counting Function

A closely related function to $p_n\#$, denoted by $z\#$, is the product of all primes less than or equal to a value z:

$$(9.10) \qquad z\# = \prod_{i=1}^{\pi(z)} p_i$$

where $\pi(z)$, the *prime counting function*, is the number of primes less than or equal to z

(9.11) $$\pi(z) = \sum_{i: p_i \le z} 1$$

Since $z\#$ only depends on primes, if the largest prime less than or equal to z is p_n (equivalently $\pi(z) = n$), then $p_n\# = z\#$. From equations (9.5) and (9.7) we can write $\sigma_0(p_n\#) = 2^n$ and

(9.12) $$\sigma_1(p_n\#) = \prod_{i=1}^{n} (p_i + 1)$$

or, expressing this in a different notation, that

(9.13) $$\sigma_1(z\#) = \prod_{i=1}^{\pi(z)} (p_i + 1)$$

We wish to show that an upper bound of $z\#$ is given by

(9.14) $$z\# \le 4^{z-1}$$

Since $\lfloor z \rfloor\# = z\#$ and $4^{\lfloor z \rfloor - 1} \le 4^{z-1}$ we can, without loss of generality, restrict z to be integer when deriving the bound. It is easy to check that (9.14) holds for small values of z. Assume then, inductively, that it holds for all values up to z. Since the value of $z\#$ only depends on primes less than or equal to z, it suffices to show that (9.14) holds for the largest odd prime less than or equal to z. Let this prime be given by $2\ell + 1$ and decompose $(2\ell + 1)\#$ into disjoint multiplications:

(9.15) $$(2\ell + 1)\# = (\ell + 1)\# \prod_{i: \ell+1 < p_i \le 2\ell+1} p_i$$

$$\le 4^{\ell} \prod_{i: \ell+1 < p_i \le 2\ell+1} p_i$$

The induction hypothesis is used to create the inequality in the first term of the last equation.

The second product term in (9.15) reminds one of a portion of a binomial coefficient. In particular, recall the computational form for binomial coefficients given in equation (2.11):

$$(9.16) \qquad \binom{2\ell + 1}{\ell} = \prod_{j=0}^{\ell-1} \frac{2\ell + 1 - j}{\ell - j}$$

There are two things to note: each term of this product is a fraction that is greater than 1 and primes in the range $i : \ell + 1 < p_i \leq 2\ell + 1$ are contained in the numerator of (9.16) but not in the denominator. This shows that the binomial coefficient is larger or equal to the multiplication of consecutive primes from $\ell + 1$ to $2\ell + 1$ and thus that

$$(9.17) \qquad \prod_{i:\ell+1<p_i\leq 2\ell+1} p_i < \binom{2\ell + 1}{\ell}$$

Since

$$\binom{2\ell + 1}{\ell} = \binom{2\ell + 1}{\ell + 1}$$

we can use the binomial summation formula of (2.21) to write

$$2\binom{2\ell + 1}{\ell} = 2^{2\ell+1} - \sum_{j\neq\ell,\ell+1} \binom{2\ell + 1}{j}$$

Dividing by 2 and eliminating the subtraction shows that

$$\binom{2\ell + 1}{\ell} < 2^{2\ell} = 4^{\ell}$$

Combining this with (9.15) yields the final bound

$$(9.18) \qquad (2\ell + 1)\,\# \leq 4^{2\ell}$$

For z not restricted to being integer or prime, this bound can be rewritten in the form of equation (9.14).

9.3   There Is Always a Prime Between n and $2n$

A generalization of a primorial number is one consisting of the multiplication of primes (not necessarily consecutive). Such numbers are said to be *square-free* since the exponents in their prime factorization are less than or equal to 1. The following proof shows that all integers can be factorized as the product of an integer and a square-free number. Let n be an integer and write

$$(9.19) \qquad\qquad n = m^2 \ell$$

where m^2 is the largest square divisor of n (possibly equal to 1) and ℓ is a square-free integer. Recall that $\omega_i(n)$ is the exponent of p_i in the prime factorization of n. Let $a_i(n)$ and $b_i(n)$ solve

$$(9.20) \qquad \omega_i(n) = 2a_i(n) + b_i(n), \quad 0 \le b_i(n) \le 1$$

If $b_i(n) = 1$, then the i'th exponent has odd parity.

With this notation, the factorization of n given by (9.19) follows for m and ℓ that satisfy $\omega_i(m) = a_i(n)$ and $\omega_i(\ell) = b_i(n)$. To illustrate this with an example, consider

$$2,156,000 = 2^5 \, 5^3 \, 7^2 \, 11$$

Then $m = 140 = 2^2 \cdot 5 \cdot 7$ and $\ell = 110 = 2 \cdot 5 \cdot 11$ and thus $2,156,000 = 140^2 \cdot 110$.

How many square-free numbers are there less than n? To answer this, note that since all products of primes less than n are square-free, the total number corresponds to the number of subsets of $\pi(n)$ items which is given by $2^{\pi(n)}$.[3] We have just showed that any number can be written as product of a square with a square-free number. There are at most $\sqrt{n}$ square numbers less than n. Thus, it must be the case that $n \le \sqrt{n} \, 2^{\pi(n)}$. After taking the natural logarithm of both sides, this implies that

$$(9.21) \qquad\qquad \pi(n) \ge \frac{\ln(n)}{2\ln(2)}$$

---

[3] A consequence of the binomial theorem, see equation (2.21).

Equation (9.21) is not only another proof of the infinitude of the primes but it also provides a lower bound for the n'th prime. You should note that this strong result is a direct consequence of the simple factorization given in (9.19).

Equation (9.9) proves useful in deriving an upper bound on the central binomial coefficient:

$$(9.22) \qquad \binom{2n}{n} = \frac{(2n)!}{(n!)^2} = \prod_{i=1}^{\infty} p_i^{\Omega_i(2n)} \Big/ \left(\prod_{i=1}^{\infty} p_i^{\Omega_i(n)} \right)^2$$

$$= \prod_{i=1}^{\infty} p_i^{\Omega_i(2n)} \Big/ \prod_{i=1}^{\infty} p_i^{2\Omega_i(n)} = \prod_{i=1}^{\infty} p_i^{\Omega_i(2n) - 2\Omega_i(n)}$$

$$= \prod_{i=1}^{\infty} p_i^{m_i}$$

where we have defined

$$(9.23) \qquad\qquad\qquad m_i = \sum_{\ell : p_i^\ell \leq 2n} \psi_{i,\ell}$$

and

$$(9.24) \qquad\qquad\qquad \psi_{i,\ell} = \left\lfloor \frac{2n}{p_i^\ell} \right\rfloor - 2 \left\lfloor \frac{n}{p_i^\ell} \right\rfloor$$

Equations (9.23) and (9.24) are the keys to calculating the upper bound. Clearly (9.24) shows that $\psi_{i,\ell} = 0$ if $p_i^\ell > 2n$. The same equation also shows that $\psi_{i,\ell}$ can at most equal 1. To establish this, observe that if x and a are positive and a integer, then[4]

$$\lfloor ax \rfloor - a \lfloor x \rfloor < a$$

Setting $x = n/p_i^\ell$ and $a = 2$ and using this inequality shows that $\psi_{i,\ell} < 2$, thus establishing the claim.

To derive the upper bound, write (9.9) in disjoint ranges as

---

[4] A quick proof goes as follows: let $x = \lfloor x \rfloor + r$, where $0 \leq r < 1$. Then the inequality follows from $a \lfloor \lfloor x \rfloor + r \rfloor = a \lfloor x \rfloor$ and $\lfloor a(\lfloor x \rfloor + r) \rfloor < a \lfloor x \rfloor + a$.

(9.25)

$$\binom{2n}{n} = \prod_{p_i \leq \sqrt{2n}} p_i^{m_{i,\ell}} \prod_{\sqrt{2n} < p_i \leq 2n/3} p_i^{m_{i,\ell}} \prod_{2n/3 < p_i \leq n} p_i^{m_{i,\ell}} \prod_{n < p_i \leq 2n} p_i^{m_{i,\ell}}$$

We now proceed to calculate a bound for each range in (9.25). There are at most $\sqrt{2n}$ primes less than or equal to $\sqrt{2n}$ and each of them is clearly less than $2n$. Thus a bound for the first range is given by

(9.26)

$$\prod_{p_i \leq \sqrt{2n}} p_i^{m_{i,\ell}} < (2n)^{\sqrt{2n}}$$

For primes that satisfy $\sqrt{2n} < p_i \leq 2n/3$ we claim that $m_i \leq 1$. Since $\psi_{i,\ell}$ can be at most 1, to show that $m_i \leq 1$ it suffices to show that $\psi_{i,2} = 0$ in this range. This follows immediately since the smallest square prime in this range is larger than $2n$. Thus, at most, this range consists of the multiplication of consecutive primes from $\sqrt{2n} + 1$ to $2n/3$. This corresponds to a primordial number and thus, using the inequality (9.14), we can write

(9.27)

$$\prod_{\sqrt{2n} < p_i \leq 2n/3} p_i^{m_{i,\ell}} \leq 4^{2n/3-1} < 4^{2n/3}$$

There are no primes in the third range: $m_i = 0$ if $2n/3 < p_i \leq n$. To show this, note that this range can be rewritten as $1 \leq n/p_i < 3/2$. Thus setting

(9.28)

$$n = p_i + r, \quad 0 \leq r < p_i/2$$

implies that $\lfloor 2n/p_i \rfloor = 2$ and $2\lfloor n/p_i \rfloor = 2$ showing that $\psi_{i,1} = 0$. To show that $\psi_{i,\ell} = 0$ for $\ell > 1$, note that (9.28) implies that

$$\frac{n}{p_i^\ell} = \frac{1}{p_i^{\ell-1}}\left(1 + \frac{r}{p_i}\right) < \frac{3}{2p_i^{\ell-1}} < 1$$

Collecting the results of (9.26) and (9.27) and substituting into (9.25) shows that

(9.29)

$$\binom{2n}{n} < (2n)^{\sqrt{2n}} 4^{2n/3} \prod_{n < p_i \leq 2n} p_i^{m_{i,\ell}}$$

Previously a lower bound was derived for the number of coin tossing games of length $2n$ that end even, see equation (6.5) . Incorporating this into (9.29):

$$\frac{4^n}{2n} < \binom{2n}{n} < (2n)^{\sqrt{2n}} 4^{2n/3} \prod_{n < p_i \leq 2n} p_i^{m_{i,\ell}}$$

uncovers the result mentioned in the beginning quote by Fine since it shows that

(9.30) $$\frac{4^{n/3}}{(2n)^{\sqrt{2n}+1}} < \prod_{n < p_i \leq 2n} p_i^{m_{i,\ell}}$$

Two quick computer programs now complete the result. The left-hand side of (9.30) increases with n and, solving it numerically, shows that it crosses 1 for $n = 468$. This guarantees that there is a prime between n and $2n$ for all $n \geq 468$. A trivial program then can be used to verify the result for values of n less than 468. Giving credit to Paul Erdős for the above analysis allows us to rephrase Fine's quote as:

> Chebyshev found them, then Paul Erdős again,
> Primes trying to hide within n and $2n$

Equivalent ways of expressing this theorem are: $p_{n+1} < 2p_n$ and $\pi(z) - \pi(z/2) \geq 1$.

There are a couple direct consequences of this result. First, it is another proof that there is no largest prime. Next, it also suggests a method to write any integer as the sum of distinct primes along with the possible addition of 1. To quickly sketch a way to construct such a sum, let p_{k_1} denote the largest prime less than or equal to n. The theorem shows that $\lfloor n/2 \rfloor + 1 \leq p_{k_1} \leq n$. If $p_{k_1} = n$, then the construction is finished. Otherwise, we are left to write $n - p_{k_1}$ as the sum of distinct primes. Again, select the largest prime less than or equal to this value and denote it by p_{k_2}. Applying the theorem again shows that $\lfloor (n - p_{k_1})/2 \rfloor + 1 \leq p_{k_2} \leq n - p_{k_1}$. If $p_{k_2} = n - p_{k_1}$, then we are done since $n = p_{k_1} + p_{k_2}$. Otherwise, the construction continues sequentially until it stops at the m'th step where either $n = p_{k_1} + \cdots + p_{k_m}$ or $n = p_{k_1} + \cdots + p_{k_m} + 1$. To illustrate the output, note that the algorithm produces the following representations: $212,506,133 = 212,506,123 + 7 + 3$ and $212,506,135 = 212,506,123 + 11 + 1$.

This construction says nothing more about the representation of an integer as the sum of primes other than constructing one.

We should mention Goldbach's conjecture, named after Christian Goldbach (1690–1764). This conjecture claims that every even integer larger than 2 can be written as the sum of two primes. This conjecture has not yet been proved (computers have not found a counter example up to about 10^{18}). The above algorithm, most frequently yields two summands, although this is clearly not mandated in its specification.

You may not have noticed that the arguments leading to equation (9.30) utilized coarse inequalities that could be substantially far from their exact values. For example, inequality (9.26) is tantamount to assuming that all positive integers less than $\sqrt{2n}$ are primes with the value $2n$ and inequality (9.27) assumes that the product of primes in the range $\sqrt{2n}$ to $2n/3$ equals the product of all primes less than n. These are extremely crude approximations to the actual values, and yet, these arguments are sufficient to establish a deep result—that a prime lies between any number and its double. How is this possible?

Let me answer the question with a question. Did you ever get an F on a test? Okay, probably not if you are reading this book. But if you did, then you would know that it is almost impossible to pass the course, and this is especially true if your F was a result of getting 0 points out of 100. This is the case for this bound. The range $2n/3$ to n grows linearly as n increases and, as proved above, there are no primes in the binomial coefficient within this range. The overestimations in the previous ranges eventually are dwarfed by the lack of primes in this range. This is the genius of the argument and shows that *even when mathematics is used as a blunt tool, it can achieve a result of fine precision.*

9.3.1   The Prime Number Theorem with a Controversy

There is a glaring hole now left in this chapter. We know that there are an infinite number of primes, know that the n'th prime must grow as $\ln(n)$ (equation (9.21)) and know that a prime always exists between a number and its double. The question left hanging concerns the asymptotic distribution of the primes among the integers. Is there a function $f(n)$, so that as n grows without bound the ratio $\pi(n)/f(n)$ converges to a non-zero value? If so, then this tell us something about the regularity of the appearance of primes in the integers. As mentioned before, there is little hope of finding a more precise answer to this question since there is no formula for

the n'th prime. In the chapter, *As Simple as 2+2=1* , we discussed
Fermat's little theorem (8.5) which can be used to test if an integer
is a prime, and Wilson's theorem, equation (8.9), which provided
necessary and sufficient conditions for an integer to be prime. These
theorems, however, do not address the distribution question. Like the
decimal digits of π which are completely deterministic but essentially
unpredictable, primes occur among the integers seemingly popping up
at random but leaving behind a madding trace of *regularity*.

The search for understanding how this series of surprises achieves a
mathematical uniformity resulted in the *prime number theorem* which
finally put the matter to rest by showing that

$$(9.31) \qquad \lim_{n\to\infty} \frac{\pi(n)}{n/\ln(n)} \to 1$$

The following table shows the value of the ratio of (9.31) converges
towards 1 for values of n from 10^1 to 10^8:

$n = 10^k$	2	4	6	8
$\pi(n)/(n/\ln(n))$	1.1513	1.1320	1.0845	1.0613

There is a long history of the discovery of this theorem, most of
which utilizes advanced mathematics. In 1948, Atle Selberg (1917–
2007) established an *elementary* approach that promised to be pivotal
in proving the result. This was achieved by both Selberg and Erdős
and resulted in a controversial interchange regarding the ownership of
the result. Even though these later techniques are elementary, they lie
outside the scope of this book.

Chapter 10
Running Off the Page

> Logical analysis is indispensable for an examination of
> the strength of a mathematical structure, but it is
> useless for its conception and design.
> The great advances in mathematics have not been
> made by logic but by creative imagination.
>
> George Frederick James Temple (1901–1992)

The analysis in this chapter illustrates Temple's observation regarding
the necessity for creative imagination in mathematics. A simple
expression is all that is needed to develop the theory of continued
fractions which leads to a deep theorem of Lagrange and also leads to
an optimal way to approximate real numbers as rational fractions.

To proceed, assume that $f(x)$ is a positive function. Obviously

$$(1 + f(x))f(x) = f(x) + f^2(x)$$

which implies that

$$(10.1) \qquad f(x) = \frac{f(x) + f^2(x)}{1 + f(x)}$$

$$= \frac{1 + f(x) + f^2(x) - 1}{1 + f(x)}$$

$$= 1 + \frac{f^2(x) - 1}{1 + f(x)}$$

This rearrangement of symbols seems like nonsense which has no
possibility of yielding a meaningful result. Before concluding this,
however, consider the case where $f(x) = \sqrt{x}$. Equation (10.1) then

© Springer Nature Switzerland AG 2020   133
R. Nelson, *A Brief Journey in Discrete Mathematics*,
https://doi.org/10.1007/978-3-030-37861-5_10

yields

(10.2) $$\sqrt{x} = 1 + \frac{x-1}{1+\sqrt{x}}$$

which expresses $\sqrt{x}$ in terms of itself. This means that substituting the right-hand side of (10.2) for the occurrence of $\sqrt{x}$ appearing in the denominator results in

$$\sqrt{x} = 1 + \cfrac{x-1}{1 + \left(1 + \cfrac{x-1}{1+\sqrt{x}}\right)} = 1 + \cfrac{x-1}{2 + \cfrac{x-1}{1+\sqrt{x}}}$$

Continuing in this way another time shows that

(10.3) $$\sqrt{x} = 1 + \cfrac{x-1}{2 + \cfrac{x-1}{1 + \left(1 + \cfrac{x-1}{1+\sqrt{x}}\right)}} = 1 + \cfrac{x-1}{2 + \cfrac{x-1}{2 + \cfrac{x-1}{1+\sqrt{x}}}}$$

and now a pattern is clear. Expressing a function in terms of itself leads to a regression. If the function appears in the denominator of the expression, then the resultant expansion is called a *continued fraction*. On first encounter, the infinite descent of fractions that threaten to run off the page appear to be inane. Changing notation can solve this run away train problem, but to show that continued fractions are anything but inane requires the rest of this chapter.

10.1   Simple Continued Fractions

A special case, termed a *simple continued fraction*, restricts all numerators to equal 1. This is denoted by

(10.4) $$[b_0, b_1, b_2, \ldots, b_n] = b_0 + \cfrac{1}{b_1 + \cfrac{1}{b_2 + \cfrac{1}{\ddots \cfrac{1}{b_n}}}}$$

With this notation, equation (10.3) with $x = 2$ shows that

(10.5) $\sqrt{2} = [1, 2, \ldots]$

To simplify notation an underline is used to represent periodic arguments. Thus, a more succinct expression of (10.5) is given by

(10.6) $\sqrt{2} = [1, \underline{2}]$

Some algebraic properties follow directly from definition (10.4) which can be summarized by the following identities:

(10.7) $[b_0, b_1, b_2, \ldots] = b_0 + \dfrac{1}{[b_1, b_2, \ldots]}$

(10.8) $[b_0, b_1, \ldots, b_{k-1}, \ b_k] = [b_0, b_1, \ldots, b_{k-2}, \ b_{k-1} + 1/b_k]$

(10.9) $[b_0, b_1, b_2, \ldots, b_{n-1}, b_n, \ldots] = [b_0, b_1, b_2, \ldots, b_{n-1}, [b_n, b_{n+1}, \ldots]]$

(10.10) $\dfrac{1}{[0, b_1, b_2, \ldots]} = [b_1, b_2, \ldots]$

To generalize the form of equation (10.6), consider the continued fraction $[a, \underline{b}]$ where a and b are non-zero integers. To derive an equation for this form, use (10.7) to write

(10.11) $[a, \underline{b}] = a + \dfrac{1}{[\underline{b}]}$

and

$$\alpha = b + \frac{1}{\alpha}$$

where $\alpha = [\underline{b}]$. This creates the quadratic equation

(10.12) $\alpha^2 = b\alpha + 1$

where only the positive solution is applicable:

$$\alpha = \frac{b + \sqrt{b^2 + 4}}{2} = -\frac{2}{b - \sqrt{b^2 + 4}}$$

Substituting this solution into (10.11) produces an equation for $[a, \underline{b}]$:

$$(10.13) \qquad [a, \underline{b}] = a - \frac{b - \sqrt{b^2 + 4}}{2} = \frac{2a - b}{2} + \frac{\sqrt{b^2 + 4}}{2}$$

Special cases of (10.13) that involve square roots are depicted in the following table:

Constant	Periodic Portion	Value
1	[2]	$\sqrt{2}$
2	[4]	$\sqrt{5}$
3	[6]	$\sqrt{10}$
4	[8]	$\sqrt{17}$
5	[10]	$\sqrt{26}$
6	[12]	$\sqrt{37}$
7	[14]	$\sqrt{50}$
8	[16]	$\sqrt{65}$
9	[18]	$\sqrt{82}$
Purely Periodic		
1	[1]	$(1 + \sqrt{5})/2$
2	[2]	$1 + \sqrt{2}$
3	[3]	$(3 + \sqrt{13})/2$
4	[4]	$2 + \sqrt{5}$
5	[5]	$(5 + \sqrt{29})/2$
6	[6]	$3 + \sqrt{10}$
7	[7]	$(7 + \sqrt{53})/2$
8	[8]	$4 + \sqrt{17}$
9	[9]	$(9 + \sqrt{85})/2$

The last section of the table deals with the special case of *purely periodic continued fractions* with unit period length. The first entry of that section of the table highlights the exquisitely beautiful example of an infinite continued fraction: that of the golden ratio[1]

---

[1]These results have been seen before, see equations (5.19) and (5.27).

(10.14) $\phi = [\underline{1}] = (1 + \sqrt{5})/2$

Before delving deeper into the patterns depicted in the above table, consider continued fractions having a periodic pattern of length of 2 given by $[a, \underline{b_0, b_1}]$. Let $\alpha = [\underline{b_0, b_1}]$ and write:

(10.15)
$$\alpha = b_0 + \frac{1}{[\underline{b_1, b_0}]} = b_0 + \frac{1}{b_1 + \dfrac{1}{\alpha}}$$

$$= b_0 + \frac{\alpha}{b_1 \alpha + 1} = \frac{(b_0 b_1 + 1)\alpha + b_0}{b_1 \alpha + 1}$$

The resultant quadratic equation

$$\alpha^2 = b_0 \alpha + b_0/b_1$$

is a generalization of (10.12) with the solution

$$\alpha = \frac{b_0 + \sqrt{b_0^2 + 4b_0/b_1}}{2} = -\frac{2b_0/b_1}{b_1 - \sqrt{b_0^2 + 4b_0/b_1}}$$

The final expression is thus given by

(10.16) $[a, \underline{b_0, b_1}] = \dfrac{2a - b_1}{2} + \dfrac{b_1 \sqrt{b_0^2 + 4b_0/b_1}}{2b_0}$

Continuing the special cases that involve square roots expands the previous table to include:

Constant	Periodic Portion	Value
1	$[\underline{1, 2}]$	$\sqrt{3}$
2	$[\underline{2, 4}]$	$\sqrt{6}$
3	$[\underline{3, 6}]$	$\sqrt{11}$
4	$[\underline{4, 8}]$	$\sqrt{18}$
5	$[\underline{5, 10}]$	$\sqrt{27}$
6	$[\underline{6, 12}]$	$\sqrt{38}$
7	$[\underline{7, 14}]$	$\sqrt{51}$
8	$[\underline{8, 16}]$	$\sqrt{66}$
9	$[\underline{9, 18}]$	$\sqrt{83}$

The path to generalizing these results to periodic sections with longer length seems clear but the necessary algebra that mimics the expansion of (10.15) soon gets out of hand. To derive a method to handle this algebra, consider the series of *convergents* given by

$$[b_0, b_1] = b_0 + \frac{1}{b_1} = \frac{b_0 b_1 + 1}{b_1}$$

$$[b_0, b_1, b_2] = b_0 + \cfrac{1}{b_1 + \cfrac{1}{b_2}} = \frac{b_0 b_1 b_2 + b_0 + b_2}{b_1 b_2 + 1}$$

$$[b_0, b_1, b_2, b_3] = \frac{b_3(b_0 b_1 b_2 + b_0 + b_2) + b_0 b_1 + 1}{b_3(b_1 b_2 + 1) + b_1}$$

To establish the general pattern for these convergents, let n_i and d_i denote the numerator and denominator of the i'th convergent of $[b_0, b_1, \ldots, b_i]$. Initial values include $n_1 = b_0 b_1 + 1$, $d_1 = b_1$ and

$$[b_0, b_1, b_2] = \frac{b_0 b_1 b_2 + b_0 + b_2}{b_1 b_2 + 1} = \frac{b_2 n_1 + b_0}{b_2 d_1 + 1} = \frac{n_2}{d_2}$$

and

$$[b_0, b_1, b_2, b_3] = \frac{b_3(b_0 b_1 b_2 + b_0 + b_2) + b_0 b_1 + 1}{b_3(b_1 b_2 + 1) + b_1} = \frac{b_3 n_2 + n_1}{b_3 d_2 + d_1} = \frac{n_3}{d_3}$$

These examples show that the numerator and denominator satisfy the general recurrence, $x_i = b_i x_{i-1} + x_{i-2}$, where each starts with different initial values

$$n_{-1} = 1, \; n_0 = b_0 \quad \text{and} \quad d_{-1} = 0, \; d_0 = 1$$

The following inductive argument will establish this recurrence relationship. Assume that the recurrence holds for all cases up to some value k. Use (10.8) to write

$$[b_0, b_1, \ldots, b_k, \; b_k + 1/b_{k+1}] = \frac{(b_k + 1/b_{k+1})n_{k-1} + n_{k-2}}{(b_k + 1/b_{k+1})d_{k-1} + d_{k-2}}$$

$$= \frac{(b_n b_{k+1} + 1)n_{k-1} + b_{k+1} n_{k-2}}{(b_k b_{k+1} + 1)d_{k-1} + b_{k+1} d_{k-2}}$$

$$= \frac{b_{k+1}(b_k n_{k-1} + n_{k-2}) + n_{k-1}}{b_{k+1}(b_k n_{k-1} + n_{k-2}) + d_{k-1}}$$

$$= \frac{b_{k+1} n_k + n_{k-1}}{b_{k+1} n_k + d_{k-1}}$$

$$= \frac{n_{k+1}}{d_{k+1}}$$

This shows the pattern continuing to the $k+1$'st case and establishes the general expression

$$(10.17) \qquad [b_0, b_1, \ldots, b_i] = \frac{n_i}{d_i} = \frac{b_i n_{i-1} + n_{i-2}}{b_i d_{i-1} + d_{i-2}}, \quad i = 0, \ldots$$

Note that, for $i = 0$, these equations yield $n_0/d_0 = b_0{}^2$.

One further relationship will prove to be useful. Define

$$\alpha_j = [b_j, b_{j+1}, \ldots]$$

and use equation (10.9) to write

$$\alpha_0 = [b_0, \ldots] = [b_0, \ldots b_i, \alpha_{i+1}], \quad i = 0, \ldots$$

Treating α_{i+1} as if it were the last part of the continued fraction permits expressing α_0 as

$$(10.18) \qquad \alpha_0 = \frac{\alpha_{i+1} n_i + n_{i-1}}{\alpha_{i+1} d_i + d_{i-1}}, \quad i = 0, \ldots$$

where (10.18) is not necessarily rational.

10.1.1 Periodic Simple Continued Fractions

To proceed with an investigation of simple continued fractions that are periodic, let α be defined by

---

[2]Later in equation (10.50) on page 152 integers n_i and d_i are shown to be co-prime.

(10.19) $\alpha = [b_0, b_1, \ldots, b_k]$

Equation (10.18) shows that

(10.20) $\alpha = \dfrac{\alpha n_k + n_{k-1}}{\alpha d_k + d_{k-1}}$

where the convergents arise from equation (10.19). Equation (10.20) corresponds to the quadratic equation

(10.21) $d_k \alpha^2 - (n_k - d_{k-1})\alpha - n_{k-1} = 0$

which has the solution

(10.22) $\alpha = \dfrac{n_k - d_{k-1} + \sqrt{(n_k - d_{k-1})^2 + 4 d_k n_{k-1}}}{2 d_k}$

$$= -\dfrac{2 n_{k-1}}{n_k - d_{k-1} - \sqrt{(n_k - d_{k-1})^2 + 4 d_k n_{k-1}}}$$

Collecting these results implies that
(10.23)

$$[a, b_0, b_1, \ldots, b_k] = \dfrac{2 n_{k-1} a - (n_k - d_{k-1})}{2 n_{k-1}} + \dfrac{\sqrt{(n_k - d_{k-1})^2 + 4 d_k n_{k-1}}}{2 n_{k-1}}$$

This equation can be used to fill out some of the square roots that were missing from the previous tables:

10.1.2   Summary of Results

The patterns depicted in the three previous tables show that there appears to be a relationship between square roots of non-square integers and continued fractions that have periodic sections from some point onward. For decimal numbers, such as $2/3 = .\overline{6}$ or $7/12 = .58\overline{3}$, cyclic patterns of this type arise if and only if the number is rational. Is there an analogous theorem for periodic continued fractions?

The entries in the tables also suggest that, if there were such a pattern, then it would pertain to irrational, rather than to rational, numbers. This follows from the fact that the square root of non-square

Constant	Periodic Portion	Value
2	$[1, 1, 1, 4]$	$\sqrt{7}$
3	$[1, 1, 1, 1, 6]$	$\sqrt{13}$
3	$[1, 2, 1, 6]$	$\sqrt{14}$
4	$[2, 1, 3, 1, 2, 8]$	$\sqrt{19}$
4	$[1, 1, 2, 1, 1, 8]$	$\sqrt{21}$
4	$[1, 2, 4, 2, 1, 8]$	$\sqrt{22}$
4	$[1, 3, 1, 8]$	$\sqrt{23}$
5	$[3, 2, 3, 10]$	$\sqrt{28}$
5	$[1, 1, 3, 5, 3, 1, 1, 10]$	$\sqrt{31}$
5	$[1, 1, 1, 10]$	$\sqrt{32}$
5	$[1, 2, 1, 10]$	$\sqrt{33}$
5	$[1, 4, 1, 10]$	$\sqrt{34}$
6	$[2, 2, 12]$	$\sqrt{41}$
6	$[1, 1, 3, 1, 5, 1, 3, 1, 1, 12]$	$\sqrt{43}$
6	$[1, 1, 1, 2, 1, 1, 1, 12]$	$\sqrt{44}$
6	$[1, 2, 2, 2, 1, 12]$	$\sqrt{45}$
6	$[1, 3, 1, 1, 2, 6, 2, 1, 1, 3, 1, 12]$	$\sqrt{46}$
6	$[1, 5, 1, 12]$	$\sqrt{47}$

integers is irrational.[3] There is also a curious form to the square root examples especially revealed in the last table: the last number in the periodic section equals twice the first number of the fraction, $b_k = 2a$, and the numbers in the periodic section (not including the last) form a palindrome: $b_i = b_{k-i-1}$, $i = 0, \ldots, k-1$. All of these observations lead us to the question: *Is something deeper at hand?*

In fact there is. These preliminary results are examples of special cases of a theorem of Lagrange that established that solutions to quadratic equations with a non-square discriminant and integer coefficients[4] have continued fractions that are periodic from some point onward. This is the analogy of the theorem for decimal expansions regarding rational numbers. Equation (10.23) is one part of Lagrange's theorem. The remaining parts will fall into place over the next few pages. The structure of the repeating portion of square roots follows

---

[3] A quick proof establishes this fact. Suppose that $\sqrt{\beta} = c/d$ for integers c and d. This implies that $d^2 \beta = c^2$. Since β is not square, there must be a prime p with an odd exponent in its factorization. All of the exponents in the prime factorizations of c^2 and d^2, however, are even. This implies that p has an odd exponent in $d^2 \beta$ and an even exponent in c^2 which means they cannot be equal. This contradicts the claim that $\sqrt{\beta}$ is rational.

[4] Such numbers are called *quadratic irrational numbers*.

as a special case of the theorem. These results are surprising and beautiful. Why would a value that is a solution to a particular type of quadratic equation impose a periodic structure on its continued fraction expansion?

There are still some results which are required to derive the necessary part of Lagrange's theorem (which was formally proved by Galois). First, a method that creates a continued fraction representation for an arbitrary number must be created. After this is completed, properties of purely periodic continued fractions are analyzed. Like periodic decimal expansions, such as $1/7 = .\overline{142857}$, purely periodic continued fractions do not have a non-periodic preamble which implies they can be written as $[\overline{b_0, b_1, \ldots, b_{\ell-1}}]$ for some period length ℓ. These investigations will then provide the apparatus needed to prove Lagrange's theorem.

10.2   General Method to Create a Continued Fraction

Let $\lfloor x \rfloor$ be the integer component of positive value x, for example $\lfloor 1.4142135 \rfloor = 1$, and let $r(x) = x - \lfloor x \rfloor$ be the decimal component of x, for example $r(1.4142135) = .4142135$. Clearly $x = \lfloor x \rfloor + r(x)$ and $r(x)$ satisfies $0 \leq r(x) < 1$. Provided that $r(x) \neq 0$ this implies that $1/r(x)$ is greater than 1 which shows that

$$(10.24) \qquad\qquad x = \lfloor x \rfloor + \frac{1}{1/r(x)}$$

We can use this equation as an operation to create a continued fraction expansion. Applying the operation on the denominator of (10.24) highlights the technique:

$$x = \lfloor x \rfloor + \frac{1}{\lfloor 1/r(x) \rfloor + 1/r(1/r(x))}$$

To recursively write the continued fraction expansion resulting from repeatedly applying (10.24), let $b_0 = \lfloor x \rfloor$. Let $\psi_0 = 1/r(x)$ and for $i \geq 1$ define

$$(10.25) \qquad\qquad b_i = \lfloor \psi_{i-1} \rfloor$$

and

(10.26) $$\psi_i = 1/r(\psi_{i-1})$$

The recursion ends when the remainder in the denominator of (10.26) equals 0. With this notation, a continued fraction expression for x can be expressed as

$$x = [b_0, b_1, \ldots, b_n]$$

where n is finite if x is a rational number (irrational numbers obviously have infinite continued fraction expansions). Checking this recursion for the golden ratio (10.14) shows that $b_0 = \lfloor \phi \rfloor = 1$ and

$$\psi_0 = \frac{1}{r(\phi)} = \frac{1}{\phi - \lfloor \phi \rfloor} = \frac{2}{\sqrt{5} - 1} = \phi$$

(the last equality was also derived in equation (5.4)). This shows that $b_1 = \lfloor \phi \rfloor = 1$. Repeating this process leads to the previously derived equation, $\phi = [\underline{1}]$.

In this example, successive iterates of equation (10.26) created a periodic sequence of unit length so that $\psi_{i+1} = \psi_i$ for all $i \geq 0$. Suppose, instead of unit length, the sequence generated a period of length ℓ so that $\psi_{i+\ell} = \psi_i$ for $i \geq 0$. Then the resultant continued fraction expansion would be purely periodic with length ℓ. To explore conditions where this occurs we next discuss properties of quadratic irrational numbers.

10.2.1 Integer Quadratics and Quadratic Surds

The quadratic equation

$$ax^2 + bx + c = 0$$

has two solutions given by

(10.27) $$r^{\pm} = \frac{-b \pm \sqrt{b^2 - 4ac}}{2a}$$

Let an *integer quadratic* be a quadratic equation where its *coefficients*, a, b and c, are integers that satisfy $a > 0$ and $b^2 + 4ac$ is not square (this last expression is termed the *discriminate*). Quadratic irrational numbers are typically represented in the form

$$(10.28) \qquad\qquad \chi^{\pm} = \frac{\alpha \pm \sqrt{\beta}}{\gamma}$$

where α and γ are rational numbers and β is a non-square integer. The value $\chi^{-} = (\alpha - \sqrt{\beta})/\gamma$ is said to be the *conjugate* of $\chi^{+} = (\alpha + \sqrt{\beta})/\gamma$. (Similarly χ^{+} is said to be the conjugate of χ^{-}.)

For every solution to an integer quadratic equation there corresponds a unique quadratic irrational number. This follows by setting $\alpha = -b$, $\gamma = a$, and $\beta = b^2 - 4ac$ which shows that $\chi^{\pm} = r^{\pm}$. Conversely, for every quadratic irrational number there corresponds a unique integer quadratic equation (this is true up to a multiplicative constant). To show this, assume that $\chi = (\alpha + \sqrt{\beta})/\gamma$ is a quadratic irrational number. Set $\alpha = -\hat{b}$ and $\gamma = 2\hat{a}$ which implies that $\hat{a} = 1/(2\gamma)$ and $\hat{b} = -\alpha/\gamma$. Equating

$$\beta = \hat{b}^2 - 4\hat{a}\hat{c} = \frac{\alpha^2 - 2\gamma\hat{c}}{\gamma^2}$$

and solving yields $\hat{c} = (\alpha^2 - \beta\gamma^2)/(2\gamma)$.

Solutions to quadratics are not altered by multiplying their coefficients by a non-zero constant. Hence we can multiply $\hat{a}$, $\hat{b}$, and $\hat{c}$ by 2γ leading to integer coefficients: $a = 1$, $b = -2\alpha$, and $c = \alpha^2 - \beta\gamma^2$. The resultant discriminant is not square since β is not square:

$$\sqrt{b^2 - 4ac} = \sqrt{4\gamma^2\beta} = 2\gamma\sqrt{\beta}$$

By construction then χ satisfies the integer quadratic $ax^2 + bx + c = 0$. Clearly the conjugate quadratic irrational number, given by $\chi' = (\alpha - \sqrt{\beta})/\gamma$, also satisfies this integer quadratic.

We note here that conjugates have algebraic properties that are best expressed by a set of identities

$$(\lambda \pm \nu)' = \lambda' \pm \nu'$$

$$(\lambda\nu)' = \lambda' \, \nu'$$

$$\left(\frac{\lambda}{\nu}\right)' = \frac{\lambda'}{\nu'}$$

$$\left(\lambda'\right)' = \lambda$$

Consider the set of numbers generated by varying α and γ of

(10.29) $$\chi = \frac{\alpha + \sqrt{\beta}}{\gamma}$$

and its conjugate

(10.30) $$\chi' = \frac{\alpha - \sqrt{\beta}}{\gamma}$$

over all integer values while keeping β constant. If β is non-square and α and γ are integer with $\gamma > 0$, then this generates an infinite set of quadratic irrational numbers. Such a group of quadratic irrational numbers *inherit their irrationality* from the same source—their common $\sqrt{\beta}$ term.

Restricting α and γ so that $\chi > 1$ and $-1 < \chi' < 0$ creates a finite set of quadratic irrational numbers which are termed *reduced quadratic surds*. To derive bounds on α and γ that satisfy these inequalities, observe that $\chi > 1$ and $\chi' > -1$ imply that $\chi + \chi' > 0$. Thus $2\alpha/\gamma > 0$ which implies that $\alpha > 0$. Since $\chi' < 0$, it must be the case that $\alpha - \sqrt{\beta} < 0$. Collecting these inequalities establishes bounds on α:

(10.31) $$0 < \alpha < \sqrt{\beta}$$

To address bounds on γ, note that $\chi > 1$ implies that $\alpha + \sqrt{\beta} > \gamma$. From $\chi' > -1$ it follows that $\alpha - \sqrt{\beta} > -\gamma$ or that $\gamma > \sqrt{\beta} - \alpha$. Collecting these inequalities produces the following bounds:

(10.32) $$\sqrt{\beta} - \alpha < \gamma < \alpha + \sqrt{\beta}$$

As an example, consider the following table that gives the set of six reduced quadratic surds, along with their associated integer quadratic equations, that inherit their irrationality from $\sqrt{7}$

α	γ	Integer Quadratic Equation	Reduced Quadratic Surds
1	2	$4x^2 - 4x - 6$	$\{(1+\sqrt{7})/2,\ (1-\sqrt{7})/2\}$
1	3	$9x^2 - 6x - 6$	$\{(1+\sqrt{7})/3,\ (1-\sqrt{7})/3\}$
2	1	$x^2 - 4x - 3$	$\{2+\sqrt{7},\ 2-\sqrt{7}\}$
2	2	$4x^2 - 8x - 3$	$\{(2+\sqrt{7})/2,\ (2-\sqrt{7})/2\}$
2	3	$9x^2 - 12x - 3$	$\{(2+\sqrt{7})/3,\ (2-\sqrt{7})/3\}$
2	4	$16x^2 - 16x - 3$	$\{(2+\sqrt{7})/4,\ (2-\sqrt{7})/4\}$

The key step in generating a continued fraction expansion is equation (10.24) which is expressed recursively with equations (10.25) and (10.26). To analyze the operation (10.24) with less awkward notion, let $e = \lfloor x \rfloor$ and $f = 1/r(x)$ and thus

$$(10.33) \qquad\qquad x = e + \frac{1}{f}$$

Assume that x is a reduced quadratic surd. Then we claim that f is also a reduced quadratic surd with the same square root value as x. To establish this, let $x = (\alpha + \sqrt{\beta})/\gamma$ and assume the associated integer quadratic equation is given by $ax^2 + bx + c = 0$ and hence $\alpha = -b$, $\gamma = 2a$ and $\beta = b^2 - 4ac$. Clearly $e + 1/f$ satisfies this integer quadratic equation so that

$$a\left(e + \frac{1}{f}\right)^2 + b\left(e + \frac{1}{f}\right) + c = 0$$

Straightforward algebra shows that

$$(10.34) \qquad (ae^2 + be + c)f^2 + (2ae + b)f + a = 0$$

Hence f is the root of this integer quadratic which can be written by

$$f = (\hat{\alpha} + \sqrt{\beta})/\hat{\gamma}$$

where $\hat{\alpha} = -(2ae + b)$ and $\hat{\gamma} = 2(ae^2 + be + c)$. Note that the discriminant of (10.34) is given by

$$\sqrt{(2ae + b)^2 - 4a(2ae + b)} = \sqrt{b^2 - 4ac} = \sqrt{\beta}$$

so that both x and f inherit their irrationality from $\sqrt{\beta}$.

To show that f is a reduced quadratic surd, first note that $0 < 1/f < 1$ since, by definition, $1/f = r(x)$. Thus $f > 1$. To form its conjugate, solve (10.33) for f

$$f = \frac{1}{x - e}$$

We can use the identities for conjugates to write

(10.35) $$f' = \frac{1}{x' - e}$$

By assumption $-1 < x' < 0$ and by definition $e \geq 1$. Thus (10.35) implies that $-1 < f' < 0$.

With this preliminary work behind us, we find ourselves at the doorway of an important result. Linking up to the recursive process (10.26), the previous discussion shows that if x is a reduced quadratic surd, then $\psi_0 = 1/r(x)$ and $\psi_i = 1/r(\psi_{i-1})$ for $i \geq 1$ are also reduced quadratic surds sharing the same square root as x. Since there are only a finite number of reduced quadratic surds with the same square root, this implies that there exists a value ℓ such that $\psi_\ell = \psi_0$ for $1 \leq \ell < \infty$. But this also implies that $\psi_{\ell+1} = \psi_1$ since $\psi_{\ell+1} = 1/r(\psi_\ell) = 1/r(\psi_0) = \psi_1$. Repeating this process shows that $\psi_{\ell+k} = \psi_k$ for all $k \geq 0$. Hence the continued fraction is purely periodic with period length ℓ. This proves that quadratic irrational numbers that are reduced quadratic surds have continued fractions that are purely periodic. Thus, for this special case, we have established Lagrange's theorem. Lagrange's theorem, however, is more general since it proves that *all* quadratic irrational numbers have a repeating structure *from some point onward*, even if they start with a non-periodic preamble. We will address this issue later in the chapter after establishing a limit property of continued fraction expansions.

An example at this point might be illustrative. Consider $\sqrt{19} = [4, \overline{2, 1, 3, 1, 2, 8}]$ which is not a reduced quadratic surd since $-1 \not< -\sqrt{19}$. Adding $4 = \lfloor \sqrt{19} \rfloor$ to this, however, produces a reduced quadratic surd which has a purely periodic continued fraction expansion: $4 + \sqrt{19} = [\overline{8, 2, 1, 3, 1, 2}]$. A simple calculation shows that there are 20 reduced quadratic surds that inherit their irrationality from $\sqrt{19}$. Six of these cycle to create the purely periodic continued fraction expansion given in the table below.

One additional fact about convergents will be all that we need to explain the special structure of continued fractions for square roots

Cycle of Reduced Quadratic Surds for the Continued Fraction of $4 + \sqrt{19}$						
Period Number	1	2	3	4	5	6
Reduced Quadratic Surd	$\frac{4+\sqrt{19}}{1}$	$\frac{4+\sqrt{19}}{3}$	$\frac{2+\sqrt{19}}{5}$	$\frac{3+\sqrt{19}}{2}$	$\frac{3+\sqrt{19}}{5}$	$\frac{2+\sqrt{19}}{3}$
Continued Fraction Digit	8	2	1	3	1	2

of non-square integers. Consider the recursion for the numerator of a convergent (equation (10.17)): $n_i = b_i n_{i-1} + n_{i-2}$. This can be rewritten as

$$(10.36) \qquad \frac{n_i}{n_{i-1}} = b_i + \frac{n_{i-2}}{n_{i-1}} = b_i + \cfrac{1}{\cfrac{n_{i-1}}{n_{i-2}}}$$

For $i = 1$ this equation shows that

$$\frac{n_1}{n_0} = b_1 + \cfrac{1}{\cfrac{n_0}{n_{-1}}} = b_1 + \frac{1}{b_0} = [b_1, b_0]$$

and for $i = 2$ that

$$\frac{n_2}{n_1} = b_2 + \cfrac{1}{\cfrac{n_1}{n_0}} = b_2 + \frac{1}{[b_1, b_0]} = [b_2, b_1, b_0]$$

The general pattern corresponds to a reversal of the digits in the continued fraction expansion. A simple induction thus shows

$$(10.37) \qquad \frac{n_i}{n_{i-1}} = [b_i, \dots, b_1, b_0]$$

Applying the same procedure for the denominator shows a similar reversal

$$(10.38) \qquad \frac{d_i}{d_{i-1}} = [b_i, \dots, b_1]$$

(the difference between (10.37) and (10.38) is a result of the different initial values).

Suppose that $x = (\alpha + \sqrt{\beta})/\gamma$ is a reduced quadratic surd so that its continued fraction expansion is given by $x = [b_0, \ldots, b_k]$ and assume its convergents are denoted by n_i/d_i. Repeating equation (10.21) shows that

(10.39) $$d_k x^2 - (n_k - d_{k-1})x - n_{k-1} = 0$$

Let y correspond to a reversal of the digits of x: $y = [b_k, \ldots, b_0]$. From the previous discussion, y has convergents given by

$$\frac{n_k}{n_{k-1}} = [b_k, \ldots, b_0] = \frac{\hat{n}_k}{\hat{d}_k}$$

and

$$\frac{d_k}{d_{k-1}} = [b_k, \ldots, b_1] = \frac{\hat{n}_{k-1}}{\hat{d}_{k-1}}$$

where $\hat{n}_{k-1} = d_k$, $\hat{d}_{k-1} = d_{k-1}$, $\hat{n}_k = n_k$, and $\hat{d}_k = n_{k-1}$. This implies that y satisfies

$$\hat{d}_k y^2 - (\hat{n}_k - \hat{d}_{k-1})y - \hat{n}_{k-1} = 0$$

or, after making the above substitutions, that

(10.40) $$n_{k-1} y^2 - (n_k - d_{k-1})y - d_k = 0$$

Setting $z = -1/y$ in (10.40) reverses the ordering of the coefficients of this quadratic leading to

(10.41) $$d_k z^2 - (n_k - d_{k-1})z - n_{k-1} = 0$$

These manipulations show that (10.41) and (10.39) are identical equations and thus x and z correspond to the quadratic's two solutions. Expressed in terms of the conjugate of x, this implies that $z = x'$

(10.42) $$x' = \frac{\alpha - \sqrt{\beta}}{\gamma} = -\frac{1}{y}$$

so that $y = -1/x'$. Thus the continued fraction of x and of y are reversals of each other.

This allows a characterization of the continued fraction expansion of $\sqrt{n}$ for non-square n. Assume that $\sqrt{n} = [c_1, c_2, \ldots]$ and note that this is not a reduced quadratic surd since $-\sqrt{n} < -1$ and thus is not purely periodic. By construction, $c_1 = \lfloor n \rfloor$ and also that $c_1 + \sqrt{n}$ is a reduced quadratic surd and thus is purely periodic

$$(10.43) \qquad x = c_1 + \sqrt{n} = [\overline{2c_1, c_2, \ldots, c_{k-1}, c_k}]$$

and in the previous paragraphs we showed that the continued fraction expansion for $y = -1/x'$ is a reverse of that for x, hence

$$(10.44) \qquad y = \frac{1}{\sqrt{n} - c_1} = [\overline{c_k, c_{k-1}, \ldots, c_2, 2c_1}]$$

Equation (10.43) implies that the form for the $\sqrt{n}$ is given by

$$(10.45) \qquad \sqrt{n} = [c_1, \overline{c_2, \ldots, c_{k-1}, c_k, 2c_1}]$$

Thus $\sqrt{n} - c_1 = [0, \overline{c_2, \ldots, c_k, 2c_1}]$ and, from relationship (10.10), that

$$(10.46) \qquad \frac{1}{\sqrt{n} - c_1} = [\overline{c_2, \ldots, c_k, 2c_1}]$$

Equations (10.44) and (10.46) represent the same continued fraction which implies that there is a palindromic relationship between the coefficients, $c_j = c_{k+2-j}$, $k = 2, \ldots, k$. Collecting these results together shows that the form of the continued fraction expansion for $\sqrt{n}$ is given by

$$(10.47) \qquad \sqrt{n} = [\lfloor n \rfloor, \overline{c_2, c_3, \ldots, c_3, c_2, 2\lfloor n \rfloor}]$$

This form is exemplified in all of the previous tables of the continued fraction expansions of square roots of non-square integers.

10.3 Approximations Using Continued Fractions

The chapter up to this point has focused on the structure of continued fraction expansions for irrational numbers and in particular concentrated on quadratic irrational numbers that have lovely expansions. This is not to say that continued fractions are not useful, however, since they are frequently used in approximations. To develop the subject along these lines it suffices to continue the example that started this chapter, that of $\sqrt{2}$. The beginning portion of its decimal expansion is given by

$$\sqrt{2} = 1.4142135623731\ldots$$

Consider a series of truncated continued fractions leading to a series of approximations. If these approximations improve as more terms are added, then the inequality

$$(10.48) \qquad \left| \sqrt{2} - [\,1,\, \underbrace{2,\ldots,2\,}_{n+1 \text{ terms}}\,] \right| < \left| \sqrt{2} - [\,\underbrace{1,2,\ldots,2\,}_{n \text{ terms}}\,] \right|, \quad n = 1,\ldots$$

should hold. For the $\sqrt{2}$ example these convergents yield

$$[1,2] = 3/2, \quad [1,2,2] = 7/5, \quad [1,2,2,2] = 17/12$$

which produces the following errors to the sequence of approximations:

$$\left| \sqrt{2} - 3/2 \right| = .085786 \ldots \qquad \left| \sqrt{2} - 7/5 \right| = .014213\ldots$$

$$\left| \sqrt{2} - 17/12 \right| = .002453\ldots$$

These first three terms corroborate the intuition that the absolute difference between the approximation and the actual result decreases as more convergents are included in the continued fraction. The following table depicts the values obtained from successive convergent approximations for $\sqrt{2}$:

Three salient features of this table pose questions which beg to be addressed. Firstly, notice the oscillation of the sign of the difference between the approximation and the exact result. Odd steps overestimate the true value where even steps underestimate it. Does a series of convergents always rotate between overshooting and undershooting the precise value?

Secondly, notice the quick convergence of the approximation to the precise value where odd steps decrease towards the true value and even steps increase towards it. Does this always occur, and if so, how can one characterize the convergence rate?

Continued Fraction Approximations of $\sqrt{2}$				
Step	b_i	n_i	d_i	$\sqrt{2} - n_i/d_i$	λ_i
1	2	3	2	$-0.0857864376269049\ldots$	2
2	2	7	5	$0.0142135623730952\ldots$	10
3	2	17	12	$-0.0024531042935716\ldots$	60
4	2	41	29	$0.0004204589248193\ldots$	348
5	2	99	70	$-0.0000721519126191\ldots$	2,030
6	2	239	169	$0.0000123789411425\ldots$	11,830
7	2	577	408	$-0.0000021239014147\ldots$	68,952
8	2	1393	985	$0.0000003644035520\ldots$	401,880
9	2	3363	2378	$-0.0000000625217744\ldots$	2,342,330
10	2	8119	5741	$0.0000000107270403\ldots$	13,652,098

Thirdly, observe that the values of n_i and d_i in the above table are always relatively prime (they have no common divisors) and the denominator steadily increases. Is this always the case?

To begin addressing these questions, consider the difference between two convergents,

$$(10.49) \qquad \frac{n_i}{d_i} - \frac{n_{i-1}}{d_{i-1}} = \frac{n_i d_{i-1} - n_{i-1} d_i}{d_{i-1} d_i}, \quad i = 2, \ldots$$

It is clear from the general recurrence equation for convergents that d_i forms an integer sequence that increases with i. Thus the denominator of (10.53), given by $d_{i-1} d_i$, is positive and increasing. Focusing on the numerator, write

$$n_i d_{i-1} - n_{i-1} d_i = (b_i n_{i-1} + n_{i-2}) d_{i-1} - n_{i-1} (b_i d_{i-1} + d_{i-2})$$

$$= b_i n_{i-1} d_{i-1} + n_{i-2} d_{i-1} - b_i n_{i-1} d_{i-1} - n_{i-1} d_{i-2}$$

$$= n_{i-2} d_{i-1} - n_{i-1} d_{i-2}$$

$$= -(n_{i-1} d_{i-2} - n_{i-2} d_{i-1})$$

Telescoping this relationship to the boundary $n_0 d_{-1} - n_{-1} d_0 = -1$ implies that

$$(10.50) \qquad\qquad n_i d_{i-1} - n_{i-1} d_i = (-1)^{i+1}$$

This shows that a linear combination of n_i and d_i equals ± 1 and answers the third question above since it implies that they must be co-prime.[5]

Defining $\lambda_i = d_{i-1}d_i$ permits rewriting (10.49) as

$$(10.51) \qquad \frac{n_i}{d_i} - \frac{n_{i-1}}{d_{i-1}} = \frac{(-1)^{i+1}}{\lambda_i}, \qquad i = 2, \ldots$$

This suggests telescoping the relationship to get the following equation:

$$\sum_{i=1}^{k} \frac{n_i}{d_i} - \frac{n_{i-1}}{d_{i-1}} = \frac{n_k}{d_k} - b_0$$

Using this, and equations (10.17) and (10.51), yields a succinct representation of a continued fraction

$$(10.52) \qquad [b_0, b_1, \ldots, b_k] = b_0 + \sum_{i=1}^{k} \frac{(-1)^{i+1}}{\lambda_i}$$

A simple recursion provides a lower bound on the rate at which λ_i increases. First note that the recursion $d_i = b_i d_{i-1} + d_{i-2}$ implies that d_i grows at least as fast as the integers. Substituting this recursion, and using the initial values of d_i, provides the following equation for λ_i:

$$(10.53) \qquad \lambda_i = \begin{cases} 0, & i = 0 \\ \\ b_i d_{i-1}^2 + \lambda_{i-1}, & i = 1, \ldots \end{cases}$$

which easily yields

$$(10.54) \qquad \lambda_i = \sum_{j=1}^{i} b_j d_{j-1}^2$$

---

[5] This follows from the fact that if they had a common multiple, so that $n_i = am$ and $d_i = bm$, then $n_i d_{i-1} - n_{i-1} d_i = m(a d_{i-1} + b n_{i-1}) = \pm 1$. This implies that m must divide 1 forcing $m = 1$.

This shows that λ_i grows at least as fast as the sum of the squared integers.

Equation (10.51) answers one portion of the first question above: truncating a continued fraction expansion creates an approximation that oscillates around a central value. Such oscillations, however, might not be around $\alpha_0 = [b_0, \ldots]$. To address this issue, use (10.18) to write the difference between α_0 and the convergents at the i'th step as:

$$(10.55) \qquad \alpha_0 - \frac{n_i}{d_i} = \frac{\alpha_{i+1}n_i + n_{i-1}}{\alpha_{i+1}d_i + d_{i-1}} - \frac{n_i}{d_i}$$

$$= \frac{d_i(\alpha_{i+1}n_i + n_{i-1}) - n_i(\alpha_{i+1}d_i + d_{i-1})}{d_i(\alpha_{i+1}d_i + d_{i-1})}$$

$$= \frac{n_{i-1}d_i - n_i d_{i-1}}{d_i(\alpha_{i+1}d_i + d_{i-1})}$$

$$= \frac{-x_i}{d_i(\alpha_{i+1}d_i + d_{i-1})}$$

$$= \frac{(-1)^i}{d_i(\alpha_{i+1}d_i + d_{i-1})}$$

This now fully answers the first question: odd and even indexed convergents successively alternate around α_0.

This brings us to the second question which can now be answered: the series of even indexed convergents increase towards α_0 whereas odd indexed convergents decrease towards it. Both reach the same limit which implies that $|\alpha_0 - n_i/d_i|$ is a decreasing sequence as conjectured in equation (10.48). These results can be summarized by an infinite series of inequalities

$$(10.56) \qquad \frac{n_0}{d_0} < \frac{n_2}{d_2} < \cdots < \alpha_0 < \cdots < \frac{n_3}{d_3} < \frac{n_1}{d_1}$$

Thus, continued fraction approximations converge to α_0 as the number of terms of the truncated continued fraction increases without bound. Observe that the last column in the table on page 152 shows a dramatic increase in the size of λ_i as i increases for the $\sqrt{2}$ example. At the tenth step, for instance, the value λ_{10} is more than 13 million and the approximation is accurate to 7 decimals.

The least that λ_i can be at each step occurs when the values of b_i are least which occurs in the case of the golden ratio where $b_i = 1$

for all i. This implies that, for a given degree of accuracy, the golden ratio requires more steps in a continued fraction approximation than any other irrational number. In this sense it is the *hardest* irrational number to approximate. Said more picturesquely, the golden ratio is the *most* irrational number! It is instructive to compare the table of its approximations to that of $\sqrt{2}$:

Continued Fraction Approximations of $\phi = (1 + \sqrt{5})/2$					
Step	b_i	n_i	d_i	$\phi - n_i/d_i$	λ_i
1	1	2	1	$-0.\,381966011250105\ldots$	1
2	1	3	2	$0.\,118033988749895\ldots$	2
3	1	5	3	$-0.\,0486326779167718\ldots$	6
4	1	8	5	$0.\,0180339887498948\ldots$	15
5	1	13	8	$-0.\,0069660112501051\ldots$	40
6	1	21	13	$0.\,0026493733652794\ldots$	104
7	1	34	21	$-0.\,00101363029772417\ldots$	273
8	1	55	34	$0.\,00038692992636546\ldots$	714
9	1	89	55	$-0.\,000147829431923263\ldots$	1,870
10	1	144	89	$0.\,000056460660007307\ldots$	4,895

Notice that at the tenth step, instead of more than 13 million with 7 digit accuracy, the value of λ_{10} is less than 5 thousand leading to an accuracy of only 4 decimals.

We have seen the pattern of numbers in this table before. From the above table it appears that $n_i = f_{i+2}$ and $d_i = f_{i+1}$. Thus, another delightful equation links the golden ratio with the Fibonacci sequence

$$(10.57) \qquad \phi \approx \frac{n_i}{d_i} = \frac{f_{i+2}}{f_{i+1}}$$

From the third property above, this also yields the immediate result that successive Fibonacci numbers are co-prime. The last column of numbers also reveal a hidden jewel since they correspond to the cumulative sum of squared Fibonacci numbers. This follows from the recurrence (10.53) and the relationship $\lambda_i = f_{i+1}f_i$ which leads to $\lambda_i = f_1^2 + \cdots + f_i^2$ (see equation (5.17) on page 70).

All of these results suggest that continued fractions are clever and accurate approximations to irrational numbers. To further support this claim, return to the portion of the denominator of equation (10.55) given by $\alpha_{i+1}d_i + d_{i-1}$. By construction, $b_{i+1} = \lfloor \alpha_{i+1} \rfloor$ and clearly $\alpha_{i+1} > 1$. This implies that

$$d_{i+1} = b_{i+1}d_i + d_{i-1} < \alpha_{i+1}d_i + d_{i-1}$$

Applying this to equation (10.55), with some minor simplifications, yields the following upper and lower bounds on the accuracy of a continued fraction approximation:

$$(10.58) \qquad\qquad |d_i\alpha_0 - n_i| < \frac{1}{d_{i+1}}$$

The form of equation (10.58) motivates a method to compare approximations. A rational approximation n/d is said to be a *best approximation* if $|d\alpha_0 - n| < \left|\hat{d}\alpha_0 - \hat{n}\right|$ for any $\hat{n}/\hat{d}$ where $n/d \neq \hat{n}/\hat{d}$ and $\hat{d} \leq d$. In this definition it is assumed that both n/d and $\hat{n}/\hat{d}$ are fractions that have been reduced to have no common factors.

10.3.1 Best Approximations

The previous results lead to a beautiful result: all best approximations are convergents from a continued fraction approximation. To prove this, first assume that $\hat{d} = d_i$ for the i'th convergent and note that the triangle inequality implies that

$$(10.59) \qquad\qquad |\hat{n} - n_i| \leq |\hat{n} - d_i\alpha_0| + |d_i\alpha_0 - n_i|$$

The bound of (10.58), and the fact that $\hat{n} \neq n_i$, implies that

$$(10.60) \qquad |\hat{n} - n_i| - |d_i\alpha_0 - n_i| > 1 - \frac{1}{d_{i+1}} = \frac{d_{i+1} - 1}{d_{i+1}} > \frac{1}{d_{i+1}}$$

The triangle inequality (10.59) thus implies that

$$\frac{1}{d_{i+1}} < |\hat{n} - n_i| - |d_i\alpha_0 - n_i| \leq |d_i\hat{n} - \alpha_0|$$

which, compared to (10.58), shows that $\hat{n}/\hat{d}$ is not a best approximation.

Assume now that $\hat{n}/\hat{d}$ is a best approximation where $\hat{d}$ is not equal to the denominator of any convergent. Without loss of generality we will prove the case where $\hat{n}/\hat{d} < \alpha_0$ (the case where $\hat{n}/\hat{d} > \alpha_0$ is completely analogous). Select i to satisfy

$$(10.61) \qquad \frac{n_{2i-2}}{d_{2i-2}} < \frac{\hat{n}}{\hat{d}} < \frac{n_{2i}}{d_{2i}} < \alpha_0 < \frac{n_{2i-1}}{d_{2i-1}}$$

All variables are integers and thus the following inequalities are satisfied:

$$(10.62) \qquad \hat{n}d_{2i-2} - n_{2i-2}\hat{d} \geq 1$$

and

$$(10.63) \qquad n_{2i}\hat{d} - \hat{n}d_{2i} \geq 1$$

Equations (10.61) and (10.51) show that

$$(10.64) \qquad \frac{\hat{n}}{\hat{d}} - \frac{n_{2i-2}}{d_{2i-2}} < \frac{n_{2i-1}}{d_{2i-1}} - \frac{n_{2i-2}}{d_{2i-2}} = \frac{1}{d_{2i-2}d_{2i-1}}$$

and

$$(10.65) \qquad \frac{n_{2i}}{d_{2i}} - \frac{\hat{n}}{\hat{d}} < \alpha_0 - \frac{\hat{n}}{\hat{d}}$$

Equation (10.62) implies that

$$\frac{\hat{n}}{\hat{d}} - \frac{n_{2i-2}}{d_{2i-2}} = \frac{\hat{n}d_{2i-2} - n_{2i-2}\hat{d}}{d_{2i-2}\hat{d}} \geq \frac{1}{d_{2i-2}\hat{d}}$$

which, with inequality (10.64), implies that

$$(10.66) \qquad \hat{d} < d_{2i-1}$$

Equation (10.63) implies that

$$\frac{n_{2i}}{d_{2i}} - \frac{\hat{n}}{\hat{d}} = \frac{n_{2i}\hat{d} - \hat{n}d_{2i}}{d_{2i}\hat{d}} \geq \frac{1}{d_{2i}\hat{d}}$$

which, with inequality (10.65), implies that

$$\frac{1}{d_{2i}} < \hat{d}\alpha_0 - \hat{n}$$

Additionally, equation (10.58) implies the following inequality

$$d_{2i-1}\alpha_0 - n_{2i-1} < \frac{1}{d_{2i}}$$

These last two inequalities thus result in

$$d_{2i-1}\alpha_0 - n_{2i-1} < \hat{d}\alpha_0 - \hat{n}$$

which, along with (10.66), contradicts the assumption that $\hat{n}/\hat{d}$ is a best approximation. The two cases above show that a best approximation cannot differ from a convergent from a continued fraction—a beautiful, and extremely useful, result.

10.4   Lagrange's Theorem and Historical Review

Before tying up some of the loose ends, let's step back in time and pay homage to some early mathematicians who somehow discovered the power of convergent approximations. The first few convergents that approximate π, for example, are given by 22/7, 333/106, and 355/113 = 3.141592. The last of these approximations, which is amazingly accurate to 6 decimals, was known to Tsu Ch'ung-Chih (429–500), a mathematician in the service of the Chinese emperor, Hsiao-wu. The same approximation was also known to Adriaan Anthonisz (1527–1607), a Dutch mathematician and surveyor. Six decimals of accuracy is achieved for the golden ratio at the 15'th convergent leading to the approximation $\phi \approx 1{,}597/987$. Similar accuracy is achieved for $\sqrt{2}$ at the 8'th convergent with the approximation $\sqrt{2} \approx 1{,}393/985$. The 7'th convergent for $\sqrt{2}$ yields 5 decimals of accuracy, $\sqrt{2} \approx 577/408$, and was known by Greek mathematicians of the fifth century B.C. as well as by Indian mathematicians of the third or fourth century B.C.

The theorem of Lagrange was left hanging in midair. Recall that the analysis only proved the special case that quadratic irrational numbers that are reduced quadratic surds have purely periodic continued fractions. The theorem states that all quadratic irrationals eventually have periodic expansions. To complete the proof, first form the conjugate of equation (10.18)

$$\alpha_0' = \frac{\alpha_{i+1}' n_i + n_{i-1}}{\alpha_{i+1}' d_i + d_{i-1}}$$

This implies that

$$\alpha_{i+1}' = -\frac{\alpha_0' d_{i-1} - n_{i-1}}{\alpha_0' d_i - n_i} = -\left(\frac{d_{i-1}}{d_i}\right)\left(\frac{\alpha_0' - n_{i-1}/d_{i-1}}{\alpha_0' - n_i/d_i}\right)$$

Previous work in the chapter shows that convergents n_{i-1}/d_{i-1} and n_i/d_i converge to α_0 as i increases without bound and that $0 < d_{i-1}/d_i < 1$ for all i. Thus, for some value $i^\star$, the value of $(\alpha_0' - n_{i^\star-1}/d_{i^\star-1})/(\alpha_0' - n_{i^\star}/d_{i^\star})$ is less than 1. It follows that this observation is also valid for all values $k \geq i^\star$ which implies that the value of α_{k+1}' from $i^\star$ onward falls between -1 and 0. The fact that $\alpha_{k+1} > 1$ for all k shows that after $i^\star$, α_{k+1} is a reduced quadratic surd. The conclusion is thus that continued fraction onward from convergent $i^\star$ is periodic. This completes the proof of Lagrange's theorem.

Appendix A
Tools of the Trade

This section lists some of the tools of the trade that are required in the book.

A.1 Recurrence Relationships

> *The secret of getting ahead is getting started.*
> *The secret of getting started is breaking your complex*
> *overwhelming tasks into small manageable tasks, and*
> *starting on the first one.*
>
> Mark Twain (1835–1910)

Consider an investment of d dollars that has a holding cost of ℓ dollars per year and yields $r > 0$ percent interest that is paid yearly. What is the amount of capital of this investment after n years? To answer this question, we follow Twain's advice above and break the problem into manageable pieces. To do this, let c_i be the capital at the end of i years for $i = 1, \ldots, n$. The statement of the problem implies that the desired answer is given by the value of c_n.

A *recursion* is set up by writing an equation for c_n in terms of previous values $\{d, c_1, \ldots, c_{n-1}\}$. In most cases, only a few previous values are needed to evaluate c_n. In particular, the value of c_n can be calculated using the previous case, c_{n-1}. To write this equation, note that the capital at the end of year n equals the capital at the end of the previous year, c_{n-1}, plus the interest gained in that year, rc_{n-1},

© Springer Nature Switzerland AG 2020
R. Nelson, *A Brief Journey in Discrete Mathematics*,
https://doi.org/10.1007/978-3-030-37861-5

less the year's holding cost, ℓ. Thus

(A.1) $$c_n = c_{n-1} + rc_{n-1} - \ell = c_{n-1}(1 + r) - \ell$$

There is nothing special about year n and thus the same form of equation holds for all years past the first:

(A.2) $$c_i = c_{i-1}(1 + r) - \ell, \quad i = 2, \ldots, n$$

The first year uses the initial capital and is given by

(A.3) $$c_1 = d(1 + r) - \ell$$

An equation like (A.1) is said to be a *recursion*.[1] After writing such an equation, the next step is to determine if there is a *closed form equation* that solves it (in our case this is an equation for c_n that includes only values d, r, and ℓ). Closed form equations can be determined in a variety of ways; one of which is to guess an answer which then is proved by induction.

To illustrate this technique, start with small values of i and *iterate* to larger values. This reveals an overall pattern:

$$
\begin{aligned}
c_2 &= c_1(1 + r) - \ell \\
&= (d(1 + r) - \ell)(1 + r) - \ell \qquad \text{substituting (A.3)} \\
&= d(1 + r)^2 - \ell(1 + r) - \ell
\end{aligned}
$$

Proceeding to c_3 reveals a similar form:

$$
\begin{aligned}
c_3 &= c_2(1 + r) - \ell \\
&= \left(d(1 + r)^2 - \ell(1 + r) - \ell\right)(1 + r) - \ell \\
&= d(1 + r)^3 - \ell(1 + r)^2 - \ell(1 + r) - \ell
\end{aligned}
$$

These equations suggest that

$$c_n = d(1 + r)^n - \ell \sum_{j=0}^{n-1}(1 + r)^j$$

---

[1] Also termed a *difference equation*.

which, using the well-known geometric summation,[2] suggests that

(A.4)
$$c_n = d(1+r)^n - \ell\frac{(1+r)^n - 1}{r}$$

Equation (A.4) can be verified to hold by induction. If a closed form solution cannot be found, then one typically solves recurrences numerically on a computer using an *iterative or recursive algorithm.*

This simple problem exemplifies a mathematical tool that is frequently used in the book (for examples, see equations (2.1), (2.3), (2.9), (2.36), (3.4), (3.16), (3.27), (7.8) and (7.10)).

A.2 Adding Zero to an Equation

Nothing Comes from Nothing

Parmenides of Elea (515 BC)

The founder of metaphysics, Parmenides of Elea (515 BC), penned the above quote which, at first, seems completely sensible. Mathematics, however, accepts nothing without proof. In fact, a commonly used technique completely ignores Parmenides' claim. This technique can be used to derive identities using the expedient of adding zero to an equation, a step which could initially seem like nonsense.

To express the technique generally, assume there is a sequence of values a_i, $i = 0, \ldots, n$, and write an equation for $a_n - a_0$ as:

$$a_n - a_0 = a_n - a_0 + (a_1 - a_1) + (a_2 - a_2) + \cdots + (a_{n-1} - a_{n-1})$$

This adds zero to the equation in the form of summands, $a_i - a_i$, $i = 0, 1, \ldots, n - 1$. The method to this madness appears when the order of the terms in the equation is changed to create

(A.5) $a_n - a_0 = a_n - a_{n-1} + a_{n-1} - a_{n-2} + \cdots + a_2 - a_1 + a_1 - a_0$

$$= \sum_{i=1}^{n} a_i - a_{i-1}$$

---

[2] See the derivation of this summation in (A.6).

The right-hand side of equation (A.5) is termed a *telescoping sum* and it is surprisingly useful in deriving a variety of identities.

To illustrate, let $a_i = x^i$ which, using (A.5), implies that

$$x^n - 1 = \sum_{i=1}^{n} x^i - x^{i-1} = (x-1) \sum_{i=1}^{n} x^{i-1}$$

Rearrangement of the terms of this equation reveals the well-known result

(A.6)
$$\sum_{i=0}^{n-1} x^i = \frac{x^n - 1}{x - 1}$$

A further identity follows immediately from this by substituting $x = y/z$ into (A.6) to get

$$\sum_{i=0}^{n-1} \left(\frac{y}{z}\right)^i = \frac{\left(\frac{y}{z}\right)^n - 1}{\frac{y}{z} - 1} = \frac{1}{z^{n-1}} \frac{y^n - z^n}{y - z}$$

This can be rewritten as

(A.7)
$$\sum_{i=0}^{n-1} y^i z^{n-i-1} = \frac{y^n - z^n}{y - z}$$

For another example, let $a_i = \binom{n}{i}$ so that

$$\binom{n}{n} - \binom{n}{0} = \sum_{i=1}^{n} \binom{n}{i} - \binom{n}{i-1}$$

The left-hand side of this equation equals 0 and the summand on the right-hand side can be simplified

$$\binom{n}{i} - \binom{n}{i-1} = \frac{n!}{(i-1)!(n-i)!} \left(\frac{1}{i} - \frac{1}{n-i+1}\right)$$

$$= \frac{n!}{(i-1)!(n-i)!} \frac{n - 2i + 1}{i(n-i+1)}$$

$$= \binom{n}{i} \left(1 - \frac{i}{n-i+1}\right)$$

The binomial theorem shows that

$$2^n = (1+1)^n = \sum_{i=0}^{n} \binom{n}{i}$$

Using both of these observations to rewrite the equation above yields

$$0 = \sum_{i=1}^{n} \binom{n}{i} \left(1 - \frac{i}{n-i+1}\right) = 2^n - 1 - \sum_{i=1}^{n} \binom{n}{i} \frac{i}{n-i+1}$$

which, with minor adjustment, yields

(A.8)
$$\sum_{i=1}^{n} \binom{n}{i} \frac{i}{n-i+1} = 2^n - 1$$

A few more examples should cinch the fact that adding zero is a useful technique to keep in the tool box. Let $a_i = \binom{2i}{i}$ with $a_0 = 1$. Elementary algebra shows that

$$a_i - a_{i-1} = \binom{2i}{i} - \binom{2(i-1)}{i-1} = \binom{2(i-1)}{i-1}\left(3 - \frac{2}{i}\right)$$

Telescoping thus implies that

(A.9)
$$\binom{2n}{n} = 1 + \sum_{k=1}^{n} \binom{2(k-1)}{k-1}\left(3 - \frac{2}{k}\right)$$

Suppose that $a_i = 1/(i+1)$. Then the telescope equation shows that

$$\frac{1}{n+1} - 1 = \sum_{i=1}^{n} \frac{1}{i+1} - \frac{1}{i} = \sum_{i=1}^{n} \frac{1}{i(i+1)}$$

which derives the identity

(A.10)
$$\frac{n}{n+1} = \sum_{i=1}^{n} \frac{1}{i(i+1)}$$

Consider setting $a_i = 1/(2^i\, i!)$ in equation (A.5) with $a_0 = 1$ and $a_i - a_{i-1} = (1 - 2i)/(2^i\, i!)$. The telescoping equation then yields the identity

(A.11)
$$\sum_{i=1}^{n} \frac{1-2i}{2^i \, i!} = \frac{1}{2^n \, n!} - 1$$

For the last example, set $a_i = 1/(i+1)!$ which yields the identity

(A.12)
$$\sum_{i=0}^{n} \frac{i}{(i+1)!} = 1 - \frac{1}{(n+1)!}$$

Creating identities in this way to some extent seems like creating something out of thin air, but this thin air often proves to be surprisingly breathable.

A.3 Induction

> Analysis and natural philosophy owe their most important discover to this fruitful means, which is called induction.
>
> Pierre-Simon Laplace (1749–1827)

Establishing that an equation is valid by induction, also termed *proof by induction*, is a mathematical technique used to demonstrate that a pattern continues indefinitely. Typically small cases are established which suggests a pattern. Next, the *induction step* is established that proves that if case k holds then case $k+1$ also holds. This establishes that the pattern continues indefinitely. Often one *guesses* the form of an equation that is satisfied by a few small cases and then uses induction to prove that the form continues for all cases. A few examples will illustrate this surprisingly powerful technique.

Consider the cumulative sum of triangular numbers $k(k+1)/2$. This sequence generates the values $1, 4, 10, 20, 35, 56$. It is easy to show that this sequence of numbers is also generated by successive values of $n(n+1)(n+2)/6$ leading to the conjecture that

(A.13)
$$\sum_{k=1}^{n} \frac{k(k+1)}{2} = \frac{n(n+1)(n+2)}{6}$$

To prove this, assume it holds up to all values n and consider the $n + 1$'st case:

$$\sum_{k=1}^{n+1} \frac{k(k+1)}{2} = \sum_{k=1}^{n} \frac{k(k+1)}{2} + \frac{(n+1)(n+2)}{2}$$

$$= \frac{n(n+1)(n+2)}{6} + \frac{(n+1)(n+2)}{2}$$

$$= \frac{1}{6}\left(n(n+1)(n+2) + 3(n+1)(n+2)\right)$$

$$= \frac{(n+1)(n+2)(n+3)}{6}$$

This shows that the pattern continues to the $n + 1$'st case and thus proves the induction step.

Another example deals with the observation that the alternating sum of squares leads to an alternating triangular sequence:

(A.14) $$\sum_{k=1}^{n}(-1)^k k^2 = (-1)^n \frac{n(n+1)}{2}$$

It is easy to see that when $n = 1$ the equation is satisfied. To prove that this pattern continues indefinitely, assume it holds for all cases up to n and consider the $n + 1$'st case:

$$\sum_{k=1}^{n+1}(-1)^k k^2 = \sum_{k=1}^{n}(-1)^k k^2 + (-1)^{n+1}(n+1)^2$$

$$= (-1)^n \frac{n(n+1)}{2} + (-1)^{n+1}(n+1)^2$$

$$= \frac{n+1}{2}\left((-1)^n n + (-1)^{n+1}2(n+1)\right)$$

$$= (-1)^{n+1}\frac{n+1}{2}\left(-n + 2(n+1)\right)$$

$$= (-1)^{n+1}\frac{(n+1)(n+2)}{2}$$

Consider the following proposed identity which is easily shown to be satisfied for small values of n:

(A.15)
$$\sum_{k=1}^{n} \frac{2k-1}{2^k k!} = 1 - \frac{1}{2^n n!}$$

This falls away effortlessly with induction

$$\sum_{k=1}^{n+1} \frac{2k-1}{2^k k!} = \frac{2(n+1)-1}{2^{n+1}(n+1)!} + \sum_{k=1}^{n} \frac{2k-1}{2^k k!}$$

$$= \frac{1}{2n!} - \frac{1}{2^{n+1}(n+1)!} + 1 - \frac{1}{2^n n!}$$

$$= 1 - \frac{1}{2^{n+1}(n+1)!}$$

This example can also be used to show that in some cases, induction and telescoping are essentially the same thing. To explain this, suppose there are functions that satisfy $g(k) - g(k-1) = f(k)$, $k = 1, \ldots, n$. Telescoping trivially creates the identity

$$g(n) = \sum_{k=1}^{n} f(k)$$

An induction argument proceeds by calculating

$$g(n+1) = \sum_{k=1}^{n+1} f(k) = f(n+1) + g(n)$$

To use identity (A.15) as an example, set

$$g(k) = 1 - \frac{1}{2^k k!}, \quad k = 1, \ldots n$$

and thus

$$f(k) = \frac{1}{2^{k-1}(k-1)!} - \frac{1}{2^k k!} = \frac{2k-1}{2^k k!}, \quad k = 1, \ldots, n$$

Adding the $n+1$'st term yields

$$\frac{2(n+1)-1}{2^{n+1}(n+1)!} + g(n) = \frac{2(n+1)-1}{2^{n+1}(n+1)!} + 1 - \frac{1}{2^n n!}$$

$$= \frac{1}{2^n n!} - \frac{1}{2^{n+1}(n+1)!} + 1 - \frac{1}{2^n n!}$$

$$= 1 - \frac{1}{2^{n+1}(n+1)!}$$

which shows that either technique could be used to derive the identity.

One more example provides convincing evidence of the power of induction. Consider the conjecture:

(A.16)
$$\sum_{k=1}^{n} kk! = (n+1)! - 1$$

It is easy to establish that this holds for $n = 1$. The induction proceeds along lines that should now be familiar

$$\sum_{k=1}^{n+1} kk! = \sum_{k=1}^{n} kk! + (n+1)(n+1)!$$

$$= (n+1)! - 1 + (n+1)(n+1)!$$

$$= (n+1)!(1 + (n+1)) - 1$$

$$= (n+2)! - 1$$

thus satisfying the induction step.

A.4 Contradiction

> *Do I contradict myself?*
> *Very Well then I contradict myself;*
> *(I am large, I contain multitudes.)*
>   Walt Whitman (1819–1892)
>   Leaves of Grass, Song of Myself

Contradiction might work in poetry but it doesn't in mathematics. The axioms that define a field of mathematics also determine the statements that can be proved within the scope of this field. Statements that can lie outside the scope of the field are neither provable nor not provable. If, however, a statement lies within the scope, then there is no other alternative that a statement be either true or false. Mathematics is the only subject that possesses such, unequivocal, certainty. One method of establishing the veracity of a mathematical statement, therefore, is to assume the opposite of the statement and show that this leads to a contradiction. This method of proof is appropriately termed, *proof by contradiction*.

A simple example is the fact that there exist an infinity of integers. To prove this, assume that there is a largest integer m. This, however, is contradicted by the fact that adding 1 to any integer increases its value. Thus, m cannot be the largest integer which establishes that integers increase without bound.

Euclid (300 BC) defined the axiomatic system of mathematics and provided a brilliant proof that there are an infinity of primes. He did this by assuming there is a largest prime, p_m, in the finite set of all prime numbers, $p_i, i = 1, \ldots, m$. From these primes, construct the integer $q = 1 + p_1 \cdot p_2 \cdots p_m$. If q is prime, then we immediately reach a contraction since $q > p_m$. Thus assume that q is not prime. If this is the case, however, then it must be divisible by a prime less than it. By the definition of q, however, it follows that q/p_i leaves a remainder of $1/q$ for all p_i thus contradicting the claim that q is not prime. We are forced to conclude that q is prime contradicting that p_m is the largest prime. Thus the primes, like the integers, increase without bound.

Another example using proof by contradiction shows that $\sqrt{5}$ is an irrational number. To prove this assume on the contrary that it is rational and can be expressed as $\sqrt{5} = a/b$ where a and b are integers and the fraction is reduced to the lowest common denominator. Squaring both sides of this equation implies that $5b^2 = a^2$. If b is even then $5b^2$ is even which implies that a^2 is even which contradicts that a/b is reduced to the lowest common denominator. Thus, b is an odd integer. Squaring a number doesn't change its odd or evenness and an odd times an odd is an odd number. This implies that $5b^2$ is odd which then implies that a is also odd. Writing $b = 2\beta + 1$ and $a = 2\alpha + 1$ leads to

$$5(4\beta^2 + 4\beta + 1) = 4\alpha^2 + 4\alpha + 1$$

which can be simplified to

$$5\beta^2 + 5\beta + 1 = \alpha^2 + \alpha$$

Factoring leads to

$$5\beta(\beta+1)+1 = \alpha(\alpha+1)$$

For any integer n, the term $n(n+1)$ is even. Thus the left-hand side of this equation is odd and the right-hand side is even which is impossible and thus contradicts the claim that $\sqrt{5}$ can be written as a rational number.

Another example shows that the sum of a rational and irrational number must be irrational. Assume then that x is rational and y is irrational and form $z = x+y$ which is claimed to be a rational number. By assumption, this means that x can be written as $x = a/b$ for integer a and b. There is no such rational representation for y but the claim is that $z = c/d$ for integer c and d. The contradiction is immediate since

$$z = \frac{c}{d} = x+y = \frac{a}{b}+y \quad\implies\quad y = \frac{c}{d}-\frac{a}{b} = \frac{cb-ad}{bd}$$

which shows that z is a rational number contradicting our assumption.

The last example of proof by contradiction involving rationality concerns a quadratic polynomial with odd coefficients. Let a, b, and c be odd numbers and consider the roots of the polynomial $ax^2+bx+c = 0$. The claim is that such roots necessarily must be irrational numbers. To prove this, assume otherwise so that a root is given by d/e with d and e being integer and reduced to have no common factor. This implies that one of values d or e is odd and the other is even. Recall that squaring a number does not change its odd or evenness and that an odd multiplied by an odd number is odd. Then

$$a(d/e)^2 + b(d/e) + c = 0 \quad\implies\quad ad^2 + bde + ce^2 = 0.$$

From the previous comments, ad^2 and ce^2 must have opposite parity and bde is even. This implies that $ad^2 + bde + ce^2$ is odd contradicting the fact that it equals the even number 0.

A.5 Order of Summations

The order of double summations is sometimes reversed to obtain a closed form solution to a summation. To review the types of interchanges found in the book define $\omega_{i,j}$, α_i, and β_j where $i,j = 1,\ldots,n$. Then common double summations include:

$$\sum_{i=1}^{n} \sum_{j=1}^{n} \omega_{i,j} \;\; = \;\; \sum_{j=1}^{n} \sum_{i=1}^{n} \omega_{i,j} \qquad \text{Independent Summations}$$

$$\sum_{i=1}^{n} \alpha_i \sum_{j=1}^{n} \beta_j \;\; = \;\; \sum_{j=1}^{n} \beta_j \sum_{i=1}^{n} \alpha_i$$

$$\sum_{i=1}^{n} \sum_{j=i}^{n} \omega_{i,j} \;\; = \;\; \sum_{j=1}^{n} \sum_{i=1}^{j} \omega_{i,j} \qquad \text{Upper Triangular}$$

$$\sum_{i=1}^{n} \alpha_i \sum_{j=i}^{n} \beta_j \;\; = \;\; \sum_{j=1}^{n} \beta_j \sum_{i=1}^{j} \alpha_i$$

$$\sum_{i=1}^{n} \sum_{j=1}^{i} \omega_{i,j} \;\; = \;\; \sum_{j=1}^{n} \sum_{i=j}^{n} \omega_{i,j} \qquad \text{Lower Triangular}$$

$$\sum_{i=1}^{n} \alpha_i \sum_{j=1}^{i} \beta_j \;\; = \;\; \sum_{j=1}^{n} \beta_j \sum_{i=j}^{n} \alpha_i$$

Appendix B
Notation and Identities Derived in the Book

Algebraic Identities		
	Equation	Page
$n = n^2 - 2\binom{n}{2}$	(4.35)	55
$n = n^3 - 3n\binom{n}{2} + 3\binom{n}{3}$	(4.35)	55
$n = n^4 - 4n^2\binom{n}{2} + 4n\binom{n}{3} + 2\binom{n}{2}^2 - 4\binom{n}{4}$	(4.35)	55
$\sum_{i=0}^{n-1} x^i = \frac{x^n-1}{x-1}$	(A.6)	164
$\sum_{i=1}^{n} \binom{n}{i}\frac{i}{n-i+1} = 2^n - 1$	(A.8)	165
$\frac{n}{n+1} = \sum_{i=1}^{n} \frac{1}{i(i+1)}$	(A.10)	165
$\sum_{i=1}^{n} \frac{1-2i}{2^i\,i!} = \frac{1}{2^n\,n!} - 1$	(A.11)	166
$\sum_{i=0}^{n} \frac{i}{(i+1)!} = 1 - \frac{1}{(n+1)!}$	(A.12)	166
$\sum_{k=1}^{n} \frac{k(k+1)}{2} = \frac{n(n+1)(n+2)}{6}$	(A.13)	166
$\sum_{k=1}^{n} (-1)^k k^2 = (-1)^n \frac{n(n+1)}{2}$	(A.14)	167
$\sum_{k=1}^{n} \frac{2k-1}{2^k k!} = 1 - \frac{1}{2^n n!}$	(A.15)	168
$\sum_{k=1}^{n} kk! = (n+1)! - 1$	(A.16)	169

© Springer Nature Switzerland AG 2020
R. Nelson, *A Brief Journey in Discrete Mathematics*,
https://doi.org/10.1007/978-3-030-37861-5

Identities involving The Golden Ratio and Fibonacci Numbers		
$\phi = \frac{1+\sqrt{5}}{2}, \;\; \psi = \frac{1-\sqrt{5}}{2}, \; f_0 = 0, \; f_1 = 1, \; f_k = f_{k-1} + f_{k-2}$	Equation	Page
$f_n = \frac{\phi^n - \psi^n}{\sqrt{5}}$	(5.10)	68
$f_{2n} = f_n(f_{n+1} + f_{n-1})$	(5.13)	69
$f_{2n} = \sum_{k=0}^{n} \binom{n}{k} f_k$	(5.14)	69
$f_n = \sum_{k=0}^{n} \binom{n}{k}(-1)^{n-k} f_{2k}$	(5.15)	70
$f_{n+1} f_{n-1} - f_n^2 = (-1)^n$	(5.16)	70
$f_i f_{i-1} = \sum_{j=1}^{i-1} f_j^2$	(5.17)	70
$f_{n+1} = \sum_{i=0}^{\lfloor n/2 \rfloor} \binom{n-i}{i}$	(5.21)	72
$f_{2n+1} = \sum_{i=0}^{n} \binom{n+i}{n-i}$	(5.22)	73

Identities involving Sum of Integer Powers		
$S_{k,n} = \sum_{i=1}^{n} i^k$	Equation	Page
$\sum_{j=1}^{k} \begin{bmatrix} k \\ j \end{bmatrix} S_{j,n} = \frac{1}{k+1} \sum_{j=1}^{k+1} \begin{bmatrix} k+1 \\ j \end{bmatrix} n^j$	(3.10)	32
$S_{1,\ell} = \begin{bmatrix} \ell+1 \\ \ell \end{bmatrix}$	(4.29)	53
$\frac{1}{2}\left(S_{1,\ell}^2 - S_{2,\ell}\right) = \begin{bmatrix} \ell+1 \\ \ell-1 \end{bmatrix}$	(4.29)	53
$\frac{1}{6}\left(S_{1,\ell}^3 - 3S_{1,\ell}S_{2,\ell} + 2S_{3,\ell}\right) = \begin{bmatrix} \ell+1 \\ \ell-2 \end{bmatrix}$	(4.29)	54
$\frac{1}{24}\left(S_{1,\ell}^4 - 6S_{1,\ell}^2 S_{2,\ell} + 8S_{1,\ell}S_{3,\ell} + 3S_{2,\ell}^2 - 6S_{4,\ell}\right) = \begin{bmatrix} \ell+1 \\ \ell-3 \end{bmatrix}$	(4.29)	54
$S_{2,\ell} = \begin{bmatrix} \ell+1 \\ \ell \end{bmatrix}^2 - 2\begin{bmatrix} \ell+1 \\ \ell-1 \end{bmatrix}$	(4.30)	54
$S_{3,\ell} = \begin{bmatrix} \ell+1 \\ \ell \end{bmatrix}^3 - 3\begin{bmatrix} \ell+1 \\ \ell \end{bmatrix}\begin{bmatrix} \ell+1 \\ \ell-1 \end{bmatrix} + 3\begin{bmatrix} \ell+1 \\ \ell-2 \end{bmatrix}$	(4.31)	54
$S_{1,n} = \frac{n(n+1)}{2}$	(7.2)	94
$S_{2,n} = \frac{n(n+1)(2n+1)}{6}$	(7.3)	96
$\sum_{\ell=0}^{k} \binom{k+1}{\ell} S_{\ell,n} = (n+1)^{k+1} - 1$	(7.5)	96
$S_{k,n} = 1 + \sum_{r=0}^{k} \binom{k}{r} S_{r,n-1}$	(7.10)	101
$S_{k,n} = \sum_{\ell=0}^{k} \binom{k}{\ell}(-1)^{k-\ell} S_{\ell,n+1}$	(7.11)	102
$\sum_{i=1}^{n-1} \sum_{j=i+1}^{n} (ij)^k = \frac{S_{k,n}^2 - S_{2k,n}}{2}$	(7.12)	102
$\sum_{\ell=0}^{k} \binom{k}{\ell} S_{k+\ell,n} = \sum_{i=1}^{n} i^k (i+1)^k$	(7.26)	105

Identities involving Triangular Numbers $T_n = \frac{(n+1)n}{2} = \binom{n+1}{2}$		
	Equation	Page
$T_n + T_{n+1} = (n+1)^2$	(7.14)	103
$T_{n+1} - T_n = n + 1$	(7.14)	103
$T_{n+1}^2 - T_n^2 = (n+1)^3$	(7.14)	103
$8T_n + 1 = (2n+1)^2$	(7.14)	103
$T_{2n+1} - T_{2n} = 2n + 1$	(7.14)	103
$T_{2n-1} - 2T_{n-1} = n^2$	(7.14)	103
$T_{n+k} = T_n + T_k + nk$	(7.15)	103
$T_{n(n+1)} = T_n + T_{n^2} + n^3$	(7.16)	103
$T_{n+2} = T_n + T_2 + 2n$	(7.16)	103
$T_{2n} = 2T_n + n^2$	(7.16)	103
$T_{T_n} = T_n + T_{T_{n-1}} + nT_{n-1}$	(7.16)	103
$T_{T_n} = T_{T_n-1} + T_n$	(7.17)	104
$T_{T_n-1} = T_{T_{n-1}} + nT_{n-1}$	(7.18)	104
$T_{2n-k} + T_{k-1} - 2T_{n-k} = n^2, \quad k = 0, \dots, n$	(7.19)	104
$T_{nk} = T_{n-1}T_{k-1} + T_n T_k$	(7.20)	104
$T_{n^2} = T_{n-1}^2 + T_n^2$	(7.21)	104
$T_{n^3} = T_{n-1}T_{n^2-1} + T_n T_{n^2}$	(7.21)	104
$T_{2T_n} = T_n \left(T_{n-1} + T_{n+1} \right)$	(7.21)	104
$T_{2n} = T_1 T_{n-1} + T_2 T_n$	(7.21)	104
$T_{2n} = 4T_n - n$	(7.22)	105
$\sum_{i=1}^{n} T_i^2 = (n+1)S_{3,n} - S_{4,n}$	(7.23)	105
$\sum_{i=1}^{n} T_i^2 = \frac{1}{4} \left(S_{4,n} + 2S_{3,n} + S_{2,n} \right)$	(7.24)	105
$\sum_{i=1}^{n} T_i^k = \frac{1}{2^k} \sum_{\ell=0}^{k} \binom{k}{\ell} S_{k+\ell,n}$	(7.25)	105
$T_i^k = \frac{1}{2^k} \sum_{\ell=0}^{k} \binom{k}{\ell} \left(1 - (-1)^{k-\ell} \right) S_{k+\ell,i}$	(7.27)	106
$n^2 T_k + nT_{k-1} = k^2 T_n + kT_{n-1}$	(7.29)	106

Identities involving Binomial Coefficients $\binom{n}{k} = \frac{n!}{k!(n-k)!}$	Equation	Page
$\binom{n}{k} = \binom{n}{n-k}$	(2.14)	10
$\binom{n}{k} = \binom{n-1}{k-1} + \binom{n-1}{k}$	(2.15)	11
$\binom{n}{k}\binom{k}{j} = \binom{n}{j}\binom{n-j}{k-j}$	(2.16)	11
$\binom{n}{k} = \frac{n}{k}\binom{n-1}{k-1}$	(2.17)	11
$\sum_{k=0}^{n}\binom{n}{k}\frac{a^k}{1+k} = \frac{(1+a)^{n+1}-1}{a(1+n)}$	(2.18)	11
$\sum_{k=0}^{n}\binom{n}{k}\frac{1}{1+k} = \frac{2^{n+1}-1}{n+1}$	(2.19)	11
$\sum_{k=0}^{n}(-1)^k\binom{n}{k}\frac{1}{1+k} = \frac{1}{n+1}$	(2.20)	11
$\sum_{k=0}^{n}\binom{n}{k}=2^n$	(2.21)	11
$\sum_{k \text{ even}}\binom{n}{k} = 2^{n-1}$	(2.22)	12
$\sum_{k \text{ odd}}\binom{n}{k} = 2^{n-1}$	(2.23)	12
$\left(1-\frac{1}{\ell}\right)^n = \sum_{k=0}^{n}\binom{n}{k}\left(\frac{-1}{\ell}\right)^k$	(2.25)	13
$(x-a)^n - (x-a)^n = 2\sum_{k \text{ odd}}\binom{n}{k}x^{n-k}a^k$	(2.26)	13
$\left(\frac{1}{x}+\frac{1}{y}\right)^n = \frac{1}{y^n}\sum_{k=0}^{n}\binom{n}{k}\left(\frac{y}{x}\right)^k$	(2.27)	13
$\sum_{k=0}^{n-i}(-1)^k\binom{n-i}{k} = \begin{cases} 0, & i=0,\ldots,n-1 \\ 1, & i=n \end{cases}$	(2.28)	14
$\sum_{k=0}^{n}\binom{n}{k}k = n2^{n-1}$	(2.29)	14
$\sum_{k=0}^{n}\binom{n}{k}k(k-1) = n(n-1)2^{n-2}$	(2.30)	14
$\sum_{k=0}^{n}\binom{n}{k}k^2 = n(n+1)2^{n-2}$	(2.31)	14
$\sum_{\ell=k}^{n}\binom{\ell}{k} = \binom{n+1}{k+1}$	(2.32)	15
$\sum_{1\le i_1 < \cdots < i_k \le n} 1 = \binom{n}{k}$	(4.36)	55
$\sum_{r=0}^{m-1}\binom{k+r}{k} = \binom{m+k}{k+1}$	(2.34)	15
$\sum_{k=0}^{n}\binom{n}{k}\frac{a^k}{k+1} = \frac{(a+1)^{n+1}-1}{(n+1)a}$	(2.18)	11
$\sum_{k=0}^{n}\binom{n}{k}(-1)^{n-k}\frac{(a+1)^{k+1}-1}{k+1} = \frac{a^{n+1}}{n+1}$	(2.55)	25

Identities involving Binomial Coefficients-Continued $\binom{n}{k} = \frac{n!}{k!(n-k)!}$	Equation	Page
$\sum_{k=0}^{n} \binom{n}{k} \frac{2^{k+1}}{k+1} = \frac{3^{n+1}-1}{n+1}$	(2.56)	25
$\sum_{k=0}^{n} \binom{n}{k}(-1)^{n-k}\frac{3^{k+1}-1}{k+1} = \frac{2^{n+1}}{n+1}$	(2.57)	25
$\sum_{k=0}^{n} \binom{n}{k}(-1)^{n-k}2^{k} = 1$	(2.51)	24
$\sum_{k=0}^{n} \binom{n}{k}(-1)^{n-k}\left(1-\frac{1}{\ell}\right)^{k} = \left(\frac{-1}{\ell}\right)^{n}$	(2.52)	24
$\sum_{k=0}^{n} \binom{n}{k}(-1)^{n-k}k2^{k-1} = n$	(2.53)	25
$\sum_{k=0}^{n} \binom{n}{k}(-1)^{n-k}k(k-1)2^{k-2} = n(n-1)$	(2.54)	25
$\sum_{k=0}^{n} \binom{n}{k}(-1)^{n-k}k(k+1)2^{k-2} = n^2$	(2.50)	24
$\sum_{k=0}^{n}(-1)^{k} \binom{n}{k}\frac{b^{k}}{k+1} = -\frac{(1-b)^{n+1}-1}{b(n+1)}$	(2.59)	26
$\sum_{k=0}^{n}(-1)^{k} \binom{n}{k}\frac{(1-b)^{k+1}-1}{b(k+1)} = -\frac{b^{n}}{(n+1)}$	(2.60)	26
$\sum_{k=0}^{n}(-1)^{k} \binom{n}{k}\frac{2^{k}}{k+1} = \begin{cases} 0, & n \text{ odd,} \\ \frac{1}{n+1}, & n \text{ even} \end{cases}$	(2.61)	26
$\sum_{k=0,\ k \text{ even}}^{n}(-1)^{k} \binom{n}{k}\frac{1}{(k+1)} = \frac{2^{n}}{n+1}$	(2.62)	26
$\binom{n}{k} = \sum_{j=0}^{n} \binom{m}{j}\binom{n-m}{k-j}, \quad 0 \le m \le n$	(2.35)	16
$\binom{2n}{n} = \sum_{k=1}^{n} \frac{2}{k}\binom{2(k-1)}{k-1}\binom{2(n-k)}{n-k}$	(6.9)	83
$\sum_{i=1}^{n} \frac{2^{2(n-i)+1}}{n}\binom{2(i-1)}{i-1} = 2^{2n} - \binom{2n}{n}$	(6.19)	87
$\binom{2n}{n} = 1 + \sum_{k=1}^{n} \binom{2(k-1)}{k-1}\left(3-\frac{2}{k}\right)$	(A.9)	165

Identities involving Binomial-R coefficients		
$\left\langle \begin{matrix} n \\ k \end{matrix} \right\rangle = \left\langle \begin{matrix} k+1 \\ n-1 \end{matrix} \right\rangle$	(2.39)	18
$\left\langle \begin{matrix} n \\ k \end{matrix} \right\rangle = \left\langle \begin{matrix} n \\ k-1 \end{matrix} \right\rangle + \left\langle \begin{matrix} n-1 \\ k \end{matrix} \right\rangle$	(2.39)	18
$\left\langle \begin{matrix} n \\ k \end{matrix} \right\rangle = \frac{n}{k}\left\langle \begin{matrix} n+1 \\ k-1 \end{matrix} \right\rangle$	(2.39)	18
$\left\langle \begin{matrix} n \\ k \end{matrix} \right\rangle = \left\langle \begin{matrix} n+1-k \\ k \end{matrix} \right\rangle$	(2.39)	18
$\sum_{r=0}^{n} \binom{k+r}{k} = \left\langle \begin{matrix} n+1 \\ k+1 \end{matrix} \right\rangle$	(2.40)	19
$\sum_{1 \le i_1 \le \cdots \le i_k \le n} 1 = \left\langle \begin{matrix} n \\ k \end{matrix} \right\rangle$	(4.42)	57
$\sum_{k=0}^{m} \left\langle \begin{matrix} n \\ k \end{matrix} \right\rangle = \left\langle \begin{matrix} n+1 \\ m \end{matrix} \right\rangle$	(2.41)	19
$\left(\frac{1}{1-x}\right)^{n} = \sum_{i=0}^{\infty} \left\langle \begin{matrix} n \\ i \end{matrix} \right\rangle x^{i}$	(2.44)	21

178

Identities involving Stirling Numbers	Equation	Page
$\left[{k+1\atop i}\right] = \left[{k\atop i-1}\right] + k\left[{k\atop i}\right]$	(3.6)	30
$n^{\overline{k}} = \sum_{i=1}^{k}\left[{k\atop i}\right]n^i$	(3.7)	30
$\sum_{j=1}^{k}\left[{k\atop j}\right]S_{j,n} = \frac{1}{k+1}\sum_{j=1}^{k+1}\left[{k+1\atop j}\right]n^j$	(3.10)	32
$n^{\overline{k}} = k\sum_{j=1}^{k-1}\left[{k-1\atop j}\right]S_{j,n}$	(3.12)	32
$n^{\underline{k}} = \sum_{i=1}^{k}(-1)^{k-i}\left[{k\atop i}\right]n^i$	(3.13)	32
$\binom{n+k}{k+1} = \frac{1}{k!}\sum_{j=1}^{k}\left[{k\atop j}\right]S_{j,n},\quad k=1,\dots,n$	(3.12)	32
$n^k = \sum_{i=1}^{k}\left\{{k\atop i}\right\}n^{\underline{i}}$	(3.17)	34
$\left\{{k\atop i}\right\} = i\left\{{k-1\atop i}\right\} + \left\{{k-1\atop i-1}\right\}$	(3.18)	34
$\sum_{i=j}^{k}(-1)^{i-j}\left\{{k\atop i}\right\}\left[{i\atop j}\right] = 0,\quad j=1,\dots,k-1$	(3.20)	35
$\left[{k\atop k-1}\right] = \left\{{k\atop k-1}\right\}$	(3.21)	35
$(p-1)!+1 \equiv_p 0$, p prime	(8.8)	116
$\sum_{i=j}^{p-1}(-1)^{p-j-1}\binom{i}{j}\left[{p-1\atop i}\right] \equiv_p 0,\quad j=1,\dots,p-2$	(8.8)	116

Identities involving Elementary Symmetric Polynomials		
$e_k(x_1,\dots,x_n)=\sum_{1\le i_1<i_2<\cdots<i_k\le n}x_{i_1}x_{i_2}\cdots x_{i_k},\quad k=1,\dots,n$ $h_k(x_1,\dots,x_n)=\sum_{1\le i_1\le i_2\le\cdots\le i_k\le n}x_{i_1}x_{i_2}\cdots x_{i_k},\quad k=1,\dots,n$ $q_k(z_1,\dots,z_n)=z_1^k+\cdots+z_n^k$ $\mathbf{1}_n=\underbrace{(1,\dots,1)}_{n\text{ times}},\ \boldsymbol{\nu}_n=(0,1,\dots,n)$	Equation	Page
$n^{\underline{k}}=\sum_{i=1}^{k}(-1)^{k-i}e_{k-i}(\boldsymbol{\nu}_{k-1})n^i$	(4.20)	50
$q_k(\boldsymbol{\nu}_{\ell-1})=S_{k,\ell-1}$	(4.26)	52
$e_{k-i}(\boldsymbol{\nu}_{k-1})=\left[{k\atop i}\right],\quad i=1,\dots,k$	(4.28)	53
$q_k(\mathbf{1}_n)=n,\quad k=1,\dots$	(4.32)	54
$e_1(\mathbf{1}_n)=n$	(4.33)	54
$e_2(\mathbf{1}_n)=\frac{1}{2}(n^2-n)$	(4.33)	54
$e_3(\mathbf{1}_n)=\frac{1}{6}(n^3-3n^2+2n)$	(4.33)	54
$e_4(\mathbf{1}_n)=\frac{1}{24}(n^4-6n^3+11n^2-6n)$	(4.33)	54
$e_k(\mathbf{1}_n)=\binom{n}{k}$	(4.34)	54
$h_k(\mathbf{1}_n)=\left\langle{n\atop k}\right\rangle$	(4.41)	57
$h_k(\boldsymbol{\nu}_n)=\left[{n+k\atop n}\right]$	(4.43)	57

$a \cup b$	union of sets a and b
$a \cap b$	intersection of sets a and b
$\|a\|$	number of elements in set a
$\mathcal{P}(a)$	power set of a
$n! = n(n-1)\cdots 1$	factorial operator
$n^{\underline{k}} = n(n-1)\cdots(n-k+1)$	falling factorial
$n^{\overline{k}} = n(n+1)\cdots(n+k-1)$	rising factorial
$\binom{n}{k} = \frac{n!}{(n-k)!\ k!}$	binomial coefficient
$\left\langle \begin{matrix} n \\ k \end{matrix} \right\rangle = \binom{n+k-1}{k}$	binomial-R coefficient
$\begin{bmatrix} k \\ i \end{bmatrix}$	Stirling number of first kind
$\begin{Bmatrix} k \\ i \end{Bmatrix}$	Stirling number of second kind
$\sum_{i=1}^{n} a_i = a_1 + \cdots + a_n$	summation operator
$\prod_{i=1}^{n} a_i = a_1 a_2 \cdots a_n$	multiplication operator
$\lfloor a \rfloor$	integer portion of a
$S_{k,n} = 1 + 2^k + \cdots + n^k$	sum of integer powers
$H_{k,n} = \sum_{i=1}^{n} \left(\frac{1}{n}\right)^k$	harmonic summation
$T_n = \frac{n(n+1)}{2}$	triangular number
$P_{k,n} = \frac{(k-2)n^2 - (k-4)n}{2}$	polygonal number
$b \equiv_n \beta$	the value of b modulus β
$\overline{b_1, \ldots, b_k} = (b_1, \ldots, b_k, b_1, \ldots, b_k, b_1 \ldots)$	repeating operator
$[b_0, \ldots]$	simple continued fraction
λ'	conjugate of a quadratic irrational number λ
$e_k(x_1, \ldots, x_n) = \sum_{1 \le i_1 < i_2 < \cdots < i_k \le n} x_{i_1} x_{i_2} \cdots x_{i_k}$	elementary symmetric polynomial
$h_k(x_1, \ldots, x_n) = \sum_{1 \le i_1 \le i_2 \le \cdots \le i_k \le n} x_{i_1} x_{i_2} \cdots x_{i_k}$	homogeneous symmetric polynomial
$q_k(z_1, \ldots, z_n) = z_1^k + \cdots + z_n^k$	symmetric power function

Bibliography

1. Martin Aigner and Günter M. Ziegler. *Proofs from THE BOOK*. Springer, Berlin, 4th ed. 2010. corr. 3rd printing 2013 edition edition, October 2009.
2. H. Davenport. *The Higher Arithmetic: An Introduction to the Theory of Numbers*. Cambridge University Press, Cambridge; New York, 8 edition, November 2008.
3. William Feller. *An Introduction to Probability Theory and Its Applications, Vol. 1, 3rd Edition*. Wiley, S.l., 3rd edition, 1968.
4. Timothy Gowers, June Barrow-Green, and Imre Leader, editors. *The Princeton Companion to Mathematics*. Princeton University Press, Princeton, third printing used edition edition, September 2008.
5. Ronald L. Graham, Oren. Patashnik, and Donald Ervin Knuth. *Concrete mathematics: a foundation for computer science*. Addison-Wesley, Reading, Mass., 1989.
6. A. Ya Khinchin. *Continued Fractions*. Dover Publications, Mineola, N.Y, revised edition, May 1997.
7. I. G. Macdonald. *Symmetric functions and Hall polynomials*. Oxford mathematical monographs. Clarendon Press ; Oxford University Press, Oxford : New York, 2nd edition, 1995.
8. Jocelyn Quaintance and H.W. Gould. *Combinatorial Identities for Stirling Numbers*. World Scientific, 5 Toh Tuck Link, Singapore 596224, 2016.

© Springer Nature Switzerland AG 2020
R. Nelson, *A Brief Journey in Discrete Mathematics*,
https://doi.org/10.1007/978-3-030-37861-5

Index

© Springer Nature Switzerland AG 2020
R. Nelson, *A Brief Journey in Discrete Mathematics*,
https://doi.org/10.1007/978-3-030-37861-5

Printed in the United States
By Bookmasters